So, You Wanna Be an
Embedded Engineer

So, You Wanna Be an Embedded Engineer

The Guide to Embedded Engineering, from Consultancy to the Corporate Ladder

by Lewin A.R.W. Edwards

AMSTERDAM • BOSTON • HEIDELBERG • LONDON
NEW YORK • OXFORD • PARIS • SAN DIEGO
SAN FRANCISCO • SINGAPORE • SYDNEY • TOKYO

Newnes is an imprint of Elsevier

Newnes

Newnes is an imprint of Elsevier
30 Corporate Drive, Suite 400, Burlington, MA 01803, USA
Linacre House, Jordan Hill, Oxford OX2 8DP, UK

 Recognizing the importance of preserving what has been written,
Elsevier prints its books on acid-free paper whenever possible.

Library of Congress Cataloging-in-Publication Data

Edwards, Lewin A. R. W.
  So, you wanna be an embedded engineer : the guide to embedded engineering,
from consultancy to the corporate ladder / Lewin A.R.W. Edwards.
      p.   cm.
  Includes index.
  ISBN-13: 978-0-7506-7953-4 (pbk. : alk. paper)
  ISBN-10: 0-7506-7953-0 (pbk. : alk. paper) 1. Embedded computer
systems—Programming—Vocational guidance. I. Title.
  TK7895.E42E378 2006
  004.16--dc22   2006015867

British Library Cataloguing-in-Publication Data
A catalogue record for this book is available from the British Library.

ISBN-13: 978-0-7506-7953-4
ISBN-10: 0-7506-7953-0

For information on all Newnes publications
visit our Web site at www.books.elsevier.com

06 07 08 09 10   10 9 8 7 6 5 4 3 2 1

Printed in the United States of America

This book is dedicated to the philosophy of making what you need out of what you can get.

Contents

1

Introduction

1.1 About This Book

Both online and in real life, nearly every day I see people asking what they need to do in order to become embedded engineers. Some are new graduates; some are still college students; a few are teenagers in high school; and a large minority are hobbyists, hardware technicians, or application-level programmers looking to improve their salary prospects and/or diversify their skills in order to avoid the 21st century plague of white-collar commoditization.

Why do so many people want to become embedded gurus? The obvious explanation is that young (and not-so-young) programmers and technicians are being lured by the glamorous, high-profile work, easy conditions, relaxed lifestyle and limitless wealth, delivered by adoring crowds, that only embedded engineering can provide. Since none of that last sentence is remotely true, however (I've been working in the field full time for somewhat more than ten years, and I don't clearly recall the last time I was pelted with cash by an adoring crowd), I can only assume that there is some major marketing campaign in progress and it is drawing people to the embedded field.

This, of course, leads to an intractable moral dilemma. Should existing embedded engineers steer these young hopefuls toward other fields, thereby keeping the pool of fresh embedded talent small, and consulting rates correspondingly lucrative? Or, should we beckon these poor innocents in the door to work on the bottom level, thereby pushing all of us embedded guys one step up the pyramid?

Humor aside, it is a generally accepted fact that the number of new graduate engineers of all types is shrinking (at least in the United States). Various theories are posited to explain this phenomenon. In the specific case of embedded

1

engineering, I see several factors causing the decline. One such factor is the unavailability in this day and age of well-documented home computers of the style my generation enjoyed—Acorn BBC, Commodore VIC-20, Sinclair ZX Spectrum, and so on. Modern personal computers are black boxes designed to run shrink-wrapped software. They are shipped without programming tools[1] and with no technical documentation whatsoever. These times in which we live are dark indeed. Operating system vendors are actively working to confine third-party software development to an exclusive club of paid-up licensees (in the same way that video game console development is controlled), and lower-level programming at the direct hardware-access layer is at best very difficult due to the unavailability of chipset documentation—in many cases, due to the manufacturers' contractual obligations to preserve trade secret information related to intellectual property protection mechanisms. In addition to this, we are faced with the mere complexity and heterogeneity of PC hardware. Back in the good old days, we could develop a homebrew program on our Commodore 64, fine-tune it to the last instruction cycle, and show it to other people with pride. This type of skill set is a vital component of embedded engineering, and it is more or less impossible to practice on mainstream home computers of the current era.

There are other factors that raise the bar or discourage people from becoming an embedded engineer, too—and I'll deal with them in the appropriate sections of this book—but the point I'm making here is that it's simply more difficult these days for kids to experiment with what they have at home; their opportunities to do so are, at best, constrained.

The good news is that a decline in the supply of engineering talent leads—inevitably—to an increased price to meet demands. Despite the specters of outsourcing and high-tech worker visas (more on these topics in Chapter 7, if you're the chapter-skipping type), right now it's a great time to be looking for work in this field. There are of course cycles and cataclysms in everything—there was an enormous crash in telecommunications engineering jobs not so long ago, for example—but over time, the trend for total employed hours and investment dollars going into high-tech engineering projects is headed relentlessly upward.

---

[1]   Apple's Mac OS is the only mainstream retail exception to this rule. While some PC vendors do offer Linux as a preload option, the people who would choose this are, in the main, people who would have installed it themselves anyway.

Note, by the way, that this book is something of a work of management heresy, almost to the point where I considered publishing it under a pseudonym. It's written for embedded engineers, or people who want to be embedded engineers, *not* people who are counting the days until the moment when they can put "manager" at the end of their job title. Although many engineers will one day deservedly graduate to management (and it's certainly not a door you should close arbitrarily), I'm assuming that right now you, dear reader, are an "individual contributor"—one of the rudely carved faces grimacing at the bottom of the totem pole. Someday I may well regret sowing these seeds of discontent in my future direct reports, but in the meantime, please make the most of the advice I'm offering.

A final note: Sprinkled throughout this book you'll find occasional humor breaks. Although most of the snippets are of my own creation, they are to a certain degree representative of engineering humor in that a fairly large percentage of engineers would find them humorous.[2] I hope that you find them funny on your first reading; if not, I encourage you to read this book again when you reach your goal of engineerhood.

1.2   What Is an Embedded Engineer?

Before I embark on a description of how you can become an embedded engineer, it is valuable to describe what the term encompasses (at least with respect to this book). To put it succinctly, embedded engineers work on the hardware and/or software of embedded control systems. In today's world, this practically always means systems built around a microprocessor core running executive control software,[3] although this software and its microprocessor aren't necessarily the main meat of the system.

Although most embedded engineers lean toward one or the other side of the software-versus-hardware developer equation, there is less rigidity to this division of labor than there is in other specialized branches of engineering. A good

---

[2]   I'll hide behind this statement any time you wince at some particularly execrable piece of levity.

[3]   For the purpose of this book, I'll consider exotic scenarios, such as pure FPGA systems (without a microcontroller core), to fall within this description.

embedded engineer is part software engineer, part digital designer, part analog designer, with at least a rudimentary understanding of radio frequency (RF)—at least as far as mitigating interference is concerned. Like a medical specialist, the embedded engineer is a general practitioner first, with additional experience that allows him or her to work with particular confidence on certain classes of problems.

This job description covers a great deal of territory—even more than you might realize from the previous broad description. A simple throwaway toy containing a speaking chip, a few LEDs, and a handful of switches; a cruise control mechanism in a car; an ultra high-speed cryptographic engine carrying a communications link between a guided missile and its operator—all these are the province of the embedded engineer. This makes it quite difficult to generalize broadly about the field without being ludicrously off-base for some significant number of readers. Perhaps more important, it makes it difficult for a newcomer who asks the question, "How do I enter the field?" to make sense of the answers that are proffered. This book tries to present several answers to many common questions, with enough background information for you to be able to decide which answer is most relevant to your skills, ambitions, and needs.

2

Education

2.1   Traditional Education Paths into Embedded Engineering

In October 2005, I attended an engineering job fair at Columbia University in New York, mainly to see who was hiring and what they were looking for. It's most instructive to do a reconnaissance mission like this occasionally, because corporations spend a large amount of money on these sorts of events and the internship and co-op programs to which they often lead. An observer can glean valuable information about the state of the industry merely from a list of who's hiring and what sort of positions they are trying to fill. I left this job fair, as I have left other similar events recently, with the following two facts in mind:

1. Domestic U.S. hiring for engineers of all types is picking up pace.

2. The traditional path of college, then internship (a polite word for apprenticeship, really), then a regular nine-to-five job with a 401(k) and dental care as the major perks is far from dead, despite what you might read about trendy companies picking résumés off the Internet and challenging job applicants to a game of foosball in order to prove their employment-worthiness.

Most of the people reading this book will probably be somewhat off the beaten path already as far as the second point is concerned. My goal is to discuss this normal path, and indicate how you can either rejoin it or work in parallel with it to the same goal.

Before you start constructing an educational plan, please be quite clear about what you're trying to achieve. Embedded work covers an unusually wide gamut of project complexities compared with most other branches of engineering. It is utterly impractical to attempt to become an all-around expert on all these things, and any prospective employer is likely to be gravely suspicious (with good reason) of such a claim, anyway. At the opposite extreme, spending ten years specializing in some arcane branch of signal processing, to the exclusion of all other knowledge, isn't a particularly useful way to spend your time either. Standards and best-practice methods evolve constantly, and design priorities shift with changes in both the supply and demand sides of the marketplace. An almost universally relevant example of this is the trend toward moving functionality—especially complex signal decomposition and waveform generation—into firmware, as digital signal processors (DSPs) and high-speed conventional microcontrollers become cheaper and more power-efficient. There is a prevailing attitude nowadays of "get these nasty analog signals into the digital domain as quickly as possible."[1] In times of yore, when micros were expensive, breathtaking (and frequently grotesque) analog, digital and hybrid front ends were assembled to avoid the need for complex firmware. Nowadays, we all need to brush up on our math skills and learn fairly complex programming techniques to work with DSPs.

The upshot of this is that you need to narrow down your goals somewhat—but not too far—before you start intensive study. It's perfectly acceptable to be somewhat unfocused while you're deciding where your interests lie—that's the only way to try new things, after all. An undergraduate degree is designed to give you some basic skills and enough generalized knowledge to use as a base to build a detailed understanding of a particular field of specialty in postgraduate study. If you're totally new to the field of engineering, I'd suggest that you structure your first year or two of undergraduate study to get all the miscellaneous mandatory subjects—mathematics, physics, chemistry, and so on—out of the way, while you converse with your professors and senior students to decide what general slice of electronics interests you the most. For example, RF engineering might appeal to you more than all the other options—in which case, you can direct your last year or two toward acquiring solid analog skills. After you earn your bachelor's degree, you can then select a field in which to specialize (preferably after experiencing this field in the real workforce).

---

[1]   "Fully digital!" has the same meaning now that "Transistorized!" had in the era of tubes (valves). It has also created approximately the same number of curmudgeons mourning the death of the old technology.

The traditional route into embedded engineering (in the United States) is a four-year bachelor's degree at an institution accredited by ABET, Inc. ABET is the official accreditor of tertiary programs in science and engineering; it is actually a consortium of twenty-eight assorted professional and technical societies. The homepage for this consortium is *<http://www.abet.org/>*, although there's not much interest there for a casual reader.

The *typical* (but far from universal) degree for embedded engineers would be a Bachelor of Science in Electrical Engineering (BSEE). Here is an example of the curriculum you can expect to cover in such a degree[2]:

Subject Matter	Credit-Hours
Chemistry	4
Physics	4
English	6
Mathematics	20
General electronics	24
Electromagnetics and electromechanics	7
Analog design	4
Digital logic and/or signal processing	12
Technical electives and projects	21
Economics or accounting	4
General courses teaching design and analysis skills	8
Liberal arts subjects for "academic rounding"	16

If you're not accustomed to the credit system used by colleges in the United States, the rule of thumb is that a "credit-hour" means one hour's classroom time per week for a semester. You can get some idea of the workload involved in all this by considering that a full-time course load is four courses per semester. This

---

[2]   This table is an amalgamation from published 2005 curricula for several U.S. colleges. I generated it by categorizing the subjects in curricula from various institutions and averaging the credit-hour weighting for each category; I then rounded it out to obtain nice integers. This information is intended *only* to be representative of the relative course load weighting of various subject matter in a typical BSEE degree.

translates to somewhere between twelve and sixteen credits. A BSEE degree is approximately eight full-time semesters. If you're willing to spend the money to attend summer and winter classes in addition to the normal spring and fall sessions, and you've got the motivation, it's possible to complete this program in somewhere between two and three years. Although you'll probably be close to insanity at the end of such an intensive study program, and when you come home, your dog will assume you're a burglar and try to bite you.

Note that there is a surprisingly wide variation in course work offered or required by BSEE degree programs at various colleges. In particular, quite a few schools are much heavier on computer science than the simulated curriculum I provided on the previous page. Degrees that mandate a large number of computer science courses (more than eight credits or thereabouts) tend to be hybrid qualifications with names such as "Bachelor of Science in Computer and Electronic Engineering." People who get these sorts of degrees might have a *very slight* head start (versus holders of standard BSEEs) working on the firmware in relatively high-end embedded systems. Regular BSEE holders might be *very slightly* advantaged in dealing with fine-tuned systems that have significant hardware issues to be solved, and perhaps a lot of hand-massaged, time-critical assembly language to tweak.[3]

This possible difference is, however, really only going to affect your first job when you get out of school.[4] After a year or two in the workforce, your useful and marketable skill set will be determined almost exclusively by the field in which you work unless you make a determined and objectively demonstrable effort at self-improvement (for example, by publishing technical articles) in some other field. There are two major reasons for this. First, and most important, once you start working in a "real" field, your learning and experience will obviously be focused into that field. You'll find that the skills you exercise in the pursuit of your day job will improve exponentially over the baseline competence level you learned in school. Skills that you aren't using will inevitably atrophy. (This will to some degree be offset by the fact that your general problem-solving abilities will increase dramatically.)

---

[3]   Note that I'm not considering whether a prospective employer will see a difference between the two qualifications.

[4]   Of course, it can affect your postgraduate study path, but I'm assuming you thought about that when you were selecting undergraduate course work.

Second, while your attention is focused on your field of choice, not only is your memory of those general college courses fading, but the practical state of the art in those other topics is moving ahead without you. For example, if you took any college-level computer science courses in the 1990s, you probably learned Pascal. If you then went away and took a ten-year sabbatical working on a farm, you'd have come back to the engineering workforce to find Pascal almost dead and buried,[5] even as a teaching language. This is perhaps a silly example, but it should illustrate to you that if you spend any significant time—two or three years at most—away from an engineering topic, you'll have to put in an appreciable effort to reclimb the learning curve when you come back to that topic.

On the topic of work experience and how to obtain it, you should give serious consideration to an internship if you're currently following the "normal" life progression from high school to college to day job. Note that by far the easiest way to get an internship is through your school. Make sure you're on the mailing lists for all the job fairs and internship opportunities. Most faculties regularly receive internship offers from corporations; these offers are sent out by email or posted on bulletin boards in the faculty buildings. Likewise, if you're on the faculty mailing list, you'll receive regular invitations to job fairs and similar events. If you're not at school, or your school for some reason doesn't attract recruiters of this sort, you can find internships directly on the websites of most large companies. To find out what's available in your area, a simple technique is to search job-related sites for local permanent positions, then go directly to the website of the employers your search uncovered and see what internships they have listed. (Although most job websites do carry listings for internships as well as regular jobs, it's been my experience that these sorts of positions are mostly advertised through direct means only.)

One or two years as an intern (even just a couple of summers will help) will not only provide the real-world experience your résumé needs in order to stand out from the pack of recent graduates, but also as a practical matter you will probably have used this time to identify your key interests and areas of strong competency. This will go a long way toward ensuring that your next job application will be for

---

[5]   No hate mail, please. I'm well aware that Delphi still exists, for example. However, Pascal's primary purpose was as a teaching language—in which role it has been almost completely overthrown, mostly by Java.

something you're good at and will enjoy. As almost a side benefit, being in the industry for a couple of years will also provide you with some networking opportunities that can be very useful when you're looking for a permanent position.

Unfortunately, if you're already in the workforce, it is probably not practical for you to cut off your regular income and go back to the internship stage. If you fall into this category, I encourage you to manufacture your own relevant experiences. The object here is to build up a portfolio of devices you've built that are relevant to a future employer looking for an embedded engineer. Pick a field that interests you on a hobby basis—model aircraft, pets, cars, laser shows—almost anything is a good example. Develop some hardware that's relevant to this field, and keep it in your portfolio. This assignment sounds almost ridiculous when stated so baldly, but if you think about it for a little while, you'll be able to see an embedded project in almost any hobby or interest. You might develop an autopilot for a model aircraft, a doggy door that uses machine vision to recognize your dog and opens only for it, an electronic ignition module for your 1960s car with its points-based ignition, and so on. By the way, local amateur radio groups are a positive hive of people who can help you learn about electronics—not just RF stuff, but all kinds of analog and digital material. Even if you don't have much interest in amateur radio for its own sake, it's worth joining one of these clubs for access to your local technical wizards. Visit the American Radio Relay League at *<http://www.arrl.org/>* to find out more about what's available in your neighborhood.

Now, if you're on the large corporation ladder, promotions beyond a certain point are generally much harder to obtain if you lack higher (postgraduate) study. You should think *very* carefully about this issue when selecting an undergraduate degree. For example, there are some schools now offering bachelor's degrees in something called *electrical engineering technology* (the qualification is called a *BSEET*). This sounds superficially like it's the right kind of course, and if you were to skim through the syllabus, you might think you see all the right buzzwords—ABET, hardware, firmware, software, programming, circuit design, and so on. However, the BSEET is a highly empirical course that is not designed to lead to further study—it's sort of the engineering equivalent of trade school. I'd advise strongly that you steer clear of this degree program; at the institutions I've examined that offer the BSEET, it is no cheaper than the BSEE, it takes approximately the same length of time to complete, and it offers significantly less in terms of employability and postgraduate study options.

One parenthetical note about postgraduate study: The utility of higher education with respect to landing a good position is not merely asymptotic; it actually has a turning point. It's a good thing to have a bachelor's degree. It's a great thing to have a master's degree. Technical certificates, industry-specific qualifications and other addenda are fine (although not usually very valuable by themselves). However, it can actually be slightly *harder* to get practical engineering jobs if you have a PhD—even to the point where some people intentionally leave them off their résumés. The stated reasons for this odd prejudice vary, but they include pay scale requirements (PhDs are expensive), perceptions about PhD holders being best suited for pure research and development positions, and the belief that PhDs are "professional students."

2.2 Getting in Without Traditional Education (and Acquiring It Thereafter)

As I started work on this chapter in early 2005, the NBC reality television series *The Apprentice* was running a contest titled "Book Smarts vs. Street Smarts." The purpose of this set of episodes was to determine (as far as televised popular entertainment can really be said to "determine" anything) whether it is college graduates or the street-savvy nondegreed who are truly best equipped to survive in the corporate world. I must confess I didn't watch the last couple of episodes, but during the time I was watching, those with street smarts were ahead by quite a large margin. This reflects a popular belief in America that a person trained in the "school of hard knocks" is more persistent, tougher and smarter than any mere bookworm. The cynic in me is quite sure that it also reflects the nondegreed target demographic of this particular genre of television programming.[6]

There is a corresponding school of thought to argue that self-taught engineers are more motivated, more flexible, and/or more creative than traditionally educated personnel. I am at best ambivalent about this viewpoint. I think it would

---

[6] An interesting and enjoyable read—though I'm not endorsing its philosophies—is *Proving You're Qualified: Strategies for Competent People Without College Degrees* by Charles D. Hayes (Autodidactic Press, June 1995, ISBN 0-9662-1201-0). There is a surprisingly large number of books covering this particular topic.

be more correct, if somewhat redundant, to say that among people without engineering-related degrees, those who are motivated enough to develop engineering skills on their own are likely to be unusually strong self-starters. This has its productivity and management advantages, but doesn't imply that these people are necessarily better engineers. Whatever the underlying justification, however, there is certainly no reason to assume that all engineering jobs are closed to you simply because you don't have a college degree in an obviously appropriate field (or indeed any degree at all). Similarly, there is almost no such thing as a degree that is completely irrelevant to embedded engineering, simply because virtually everything in our daily existence is controlled or at least affected by embedded control systems.

Before I go on, however, you should note that I don't recommend you take any of these nonstandard paths by choice. I'm providing the advice in this chapter because I realize—having lived through this exact path myself—that it sometimes happens that you migrate into a full-time career before you have a chance to obtain the "prerequisite" formal qualifications. A large number of people do pursue very successful and lucrative embedded engineering careers all across America without ever obtaining a degree in this field—or any other, for that matter. If you want to emulate their example, it's certainly possible. However, I strongly advise that you enroll in a suitable degree program at the earliest opportunity; it is the path of least resistance for future advancement and will definitely make your life considerably easier, besides increasing your earning potential.

Now, if you lack the standard BSEE or equivalent qualification, your employment options will be constrained but not entirely extinguished. You will, however, have to put in significantly more effort than your traditionally educated colleagues to get a good position. How do you start?

To begin with, I would advise you not even to bother submitting your résumé for positions advertised on job-related websites unless you have *exactly* the qualifications, experience, and other requirements spelled out in the advertisements. There are two reasons for this: first, these job postings attract a staggering number of applicants and your details are very likely to be lost in the noise. More important however, as of February 2006, new and exceedingly irritating "promotion of diversity" legislation went into effect in the United States.[7] The net effect of this legislation is that you will need to tailor your résumé to the job for which you're applying, and make sure that you address each posted requirement point for point. If the job requires three years of telecoms experience and you have two, your résumé will be tossed, even if your two years of experience involved completely redesigning and singlehandedly rewiring the telecoms infrastructure of a small European nation using only your teeth and a roll of twine, and you were awarded the Nobel Prize for your efforts. (Yes, this system is fully as insane as it sounds. It will naturally result in a mediocre applicant who checks the right boxes on the checklist winning out over the person who's best for the job. As if Human Resources *needed* yet another utterly arbitrary reason to take applicants out of the candidate pool!)

Note that the aforementioned bureaucratic irritation will only affect you when applying for jobs at mid-size to large companies (more than 50 employees). This fact alone is a good reason to focus your search on smaller companies. However, you might prefer to direct your search in this direction anyway, because smaller companies will practically always pay *much* more attention to your experience, your portfolio, and other tangibles than to your formal qualifications. I discuss life in a small company in more detail in Chapter 6, but by way of summary: in a small company, you're constantly (from first interview to retirement) going to be dealing at a close face-to-face level with the principals in charge of hiring and firing. Hence, you have a much better opportunity in the small-company interview to impress someone with *immediate* decision-making power. Make a good impression, and you can be hired more or less on the spot because the interviewer doesn't need to refer the decision upward and/or to justify it to any other stakeholder.

In contrast, during the large-company hiring process, you need to satisfy a chain of people. HR's paperwork checklist (or even an automated software process; see the following) comes first, then you need to impress the hiring manager, and finally you need to provide enough satisfying information for the hiring manager to write up a convincing argument for upper management as to why you should be hired. Plus, if all that wasn't enough, you will probably have to face many more

---

[7]   A good summary of the situation was published in Anne Fisher's CNNmoney column under the title "Job hunting online gets trickier"; at the time of this writing, it is viewable at *<http://money.cnn.com/2006/02/06/news/economy/annie/annie_0206/index.htm>*.

competitors, if only because jobs in large companies are advertised more widely than small-company positions. Of course, you may feel that the healthcare, tuition reimbursement, 401(k) plan, stock options, in-chair massages, and other perks of the big company[8] are absolutely necessary to your happiness at work, in which case you don't have a choice. I'm not saying it's impossible for a nondegreed or at least nontraditionally educated engineer to get a Fortune 500 job—I managed it twice, in fact—but it's not easy. You'll have to show that you bring some very unusual talent to the party in order to stay in the running for the position.

Regardless of whether you aim for a position with a large or a small company, it is my experience that the best way to get a good placement is through a recruiter who will work on your behalf from inside the target employer. This is particularly true for people with a nontraditional educational background. I have obtained by far the best results through small recruiters who service relatively few corporate clients and give a high level of personal attention to these clients and their prospective candidates. The large national clearinghouse type recruiters make heavy use of automatic filtering software to match résumés against positions, with exceedingly poor results overall. If I may be permitted the luxury of a sweeping generalization, the boiler room recruiters who work at these big recruitment agencies are also utterly ignorant of the technical details of any of the industries they represent. Consequently, they are almost completely incapable of intelligently matching résumé bullet points to the requirements of a particular job vacancy.

One technique I have used myself, and recommended with good results to others, is to search the usual online job databases[9]—not for positions per se, but for recruiters. The majority of posted openings are listed by third-party recruiters, not directly by the hiring companies. These listings generally include the recruiter's contact details. Instead of using the automatic "Apply for this job now!" button on the website, contact the recruiter directly—by telephone is generally best, with an email follow-up—and provide your résumé. State which job you were looking at, but specify that you would like your résumé to be placed on file for consideration for other positions.

---

[8]   If you find a company still offering all this stuff, please drop me a line.
[9]   HotJobs, Monster, BrassRing, CyberCoders, and so on—there are literally hundreds of employment websites that cater partially or entirely to high-tech positions.

While you're visiting these job sites, of course it's also worth posting your résumé as publicly searchable—I've received some excellent employment leads from recruiters who trawl the online candidates' databases looking for talent. Don't be too disappointed if you don't get anything worthwhile by this route, though—it's very hit or miss.

I've been asked, on occasion, if it is a good investment to pay for subscription-type employer referral agencies and services that, for a fee, submit your résumé for job openings in numerous different forums. My opinion is that these services are an unmitigated waste of money for the average engineer, and this is doubly true for anyone who doesn't have a normal educational background. These services simply take your résumé, and use automatic software to submit it to dozens of different jobs on websites. This is in my view the least efficient, least likely to succeed method you could possibly use to look for a job. Your first clue to this should be the fact that these agencies advertise primarily by means of spam email.

By the way, the algorithmic candidate-to-position matching software systems I mentioned earlier, although doubtless operating precisely as designed, are utterly incapable of placing employees with nonstandard educational histories. This is because they almost invariably give filter priority to formal qualifications. For instance, while I was employed as a design engineer at one large U.S. company, I regularly received automated emails from their recruiting system offering me positions at a repair technician level *working in Iraq* (at less than half my then-current salary, even counting the danger money bonus for working in a Middle Eastern war zone). This sort of behavior is utterly typical of automated recruitment software.

Working for a pittance repairing electronic equipment among the sands of the Middle East is possibly something of an extreme case. However, you might find it necessary to take a less-than-ideal position, or a less-than-hoped-for salary point, in order to reach your ultimate employment goal. Even if you've managed to convince the hiring manager that you will make a perfect addition to the team, they still have to justify the decision upward. It's easier for the manager to hire you at a lower level than you deserve and then ensure you're promoted or get a salary raise later, once you have proved your worth to the company. So, I'd advise you to be willing to take something less than you actually want, with the view to being rewarded for short-term performance.

This does put the onus on you for two things. First, you must obviously be capable of demonstrating good performance within a reasonable review period—say, six months. The duration of this review period should be something you negotiate explicitly with the manager during the interview process. Second, you must be willing to initiate contact with the manager after the evaluation period expires. This doesn't need to be confrontational; you are merely reminding them that you mutually agreed to review your performance after a given time period, which has now elapsed. By the way, I suggest you keep at least a weekly log of what you get done during this time; it will help you assemble a strong argument when it comes to review time.

Once you score the plush engineering job of your dreams, you should *immediately* begin planning for your future career path and ensuring that you keep your options as wide open as possible. If you're in a large company, you'll need to normalize your educational background in order to increase your job security and progress up the management ziggurat.[10] If you're in a smaller company, tidying up your credentials will help you if the company falls on hard times or is acquired and you have to search for a new position. The bottom line here: Given that you have a modicum of financial stability in a job that is earning you valuable "experience points" in the field where you want to work, I strongly advise that you invest the time and effort to complete an engineering-related degree.

A brief word is due at this point about undergraduate study for those who didn't recently exit high school. If you fall in the target demographic I had in mind for this book, you are probably already in mid-career, albeit probably not as an engineer. This means that you qualify as what the universities used to call a "mature-age" student. (This term has definitely fallen into some disfavor, but no clear replacement has emerged.) You may therefore have some trepidation about embarking on a new course of undergraduate study, and I'd like to dispel this feeling and provide a few real-world experiences to help you with this step.

The first thing you need to do is simply overcome the inertia required to apply. Miscellaneous tasks that fall into this heading are choosing an institution,

---

[10]   This is true even if you don't specifically desire to be a manager. See Chapter 6 for more information on why this is the case.

obtaining your high school (and college, if any) transcripts, vaccination records, and filling in the application form. Actually getting in is not likely to be a problem unless you set your sights really high when you were choosing a college. The experience I've had in my own endeavors, and when discussing the matter with others similarly placed, is that it's substantially easier to get into college as a mature-age student than it is for a high school graduate. At worst, you might have to do a semester or two as a nonmatriculated student, in order to establish a respectable grade point average. Again, my own experience, shared with several of my colleagues, is that practically all I had to do was write a convincing 500-word essay describing what I had been doing in my day job for the past few years. I also pointed out that my employer's reimbursement policy was contingent on good results—thereby implying to the college admissions department that I would be highly motivated to achieve excellent grades.

You also need to consider how you're going to pay for this degree.[11] Tertiary education in the United States is flabbergastingly expensive. Tuition fees for four-year degree programs (excluding books and other supplies, and of course not including accommodations and other optional extras either) fall into essentially three tiers: approximately $5,000 per full-time year for a public institution, $21,000 per year for a mid-range private college, $40,000 per year and upward for an Ivy League school. Unquestionably, your first port of call to pay for this should be your employer—most large U.S. companies and many small ones offer at least some educational assistance, and many larger companies will pay the entire cost of your degree, often even including books and other supplies.[12] All you have to pay out of your own pocket is tax on this additional bonus income, and a certain amount will be tax-free depending on your income level and eligibility for certain credits.

If the employer reimbursement avenue is not available, the next best option is to work out a pace of study that will allow you to pay for school with a portion

---

[11] I intentionally put this point after the "choose a college and apply" stage, because the pure-earnings return on investment for a good degree is so enormous (compared to the risk)—easily the best *guaranteed* investment you'll ever make, barring a lucky break in real estate—that the tuition price of a college should not be the primary reason for selection.

[12] This is yet another excellent reason why you should go back to school. In this age of dwindling or nonexistent bonuses and stock options, you can potentially write yourself what amounts to a bonus check of $10,000 or even more per year just by taking advantage of your employer's reimbursement plan. Frankly, it is insanity *not* to grab large fistfuls of this free money, if offered the opportunity.

of your regular salary (as opposed to using cash out of your savings account). Depending on where you are in your career path (earlier is better), you might consider diverting contributions out of your 401(k) or other retirement fund into your education program. Not all financial advisers would agree with that suggestion, by the way—the reason I believe it's sound advice is that the effective return on investment for your first degree is quite substantial. A frequently quoted statistic from the Census Bureau states that the average lifetime income for a high school graduate is $1.2 million; the average lifetime income for holders of bachelors' degrees is $2.1 million. Assuming the degree costs you $80,000, that's an APY of approximately 5.5%,[13] which is respectable but not amazing.

To get to the amazing part, I'd like you to observe that those crude figures are averaged across all professions and locations. If we consider a 45-year working lifespan, the Census Bureau figures average out to $26,667 per year for the high school graduate, or $46,667 per year for the college graduate. Now, $45,000 or thereabouts is an average ballpark *starting* salary for a recent graduate electronic engineer in most parts of the United States.[14] Assuming a regular annual raise of 3.5%,[15] if you were to get an engineering job right out of college, you'll be above that $46,667 baseline after just two years. Over the nominal 45-year lifespan, you would actually earn $4.2 million. At that rate, the effective APY on your investment is around 8.4%, which is significant by anyone's measure. It can get even better than that: if you consider the ideal case where your employer pays for the degree (let's assume that you're in a 35% tax bracket), you'll actually only pay something like $20,000 in taxes[16] for the degree.

For an entertaining discussion of the cost of tertiary education and how you should prioritize it, as well as numerous financial tips that can be useful to

---

[13] Over a 45-year working life.

[14] I obtained this figure by averaging salary.com query data for various ZIP codes, looking for entry-level embedded engineering titles.

[15] This doesn't necessarily represent an actual raise, of course; it's merely (hopefully) keeping pace with cost-of-living increases. A 3.5% raise happens to be the "industry average" of what is issued by large corporations for engineering-type professions in 2005. In private industry, 4.5% to 5.5% is probably a better figure to shoot for. One-time bonuses aren't included in this figure.

[16] I'm well aware that $20,000 is not 35% of $80,000. It's very hard to estimate the actual post-tax cost of a reimbursed degree sensibly because it depends, among other things, on how quickly you study—there's a fixed dollar amount of tuition expenses you can deduct each tax year. If you restrict yourself to spending only this amount, you can actually get your employer-sponsored degree totally tax free, but it will take something like sixteen years of study!

people making the kind of career switches I'm discussing in this book, you may enjoy reading *The Money Book for the Young, Fabulous & Broke* by Suze Orman (Riverhead Hardcover, March 2005, ISBN 1-5732-2297-6). Note that there is a paperback edition as well, but it doesn't seem to be cataloged properly; the book catalogs I have checked say it's due for release in 2025!

Depending on your immigration status, income level and other eligibility criteria, you should also investigate the possibility of federal grants and student loans. Of course, it goes without saying that you should apply for any appropriate scholarships, if available. It's always worth asking your school's bursar and/or a course adviser about scholarship options at the end of every semester, particularly if you're getting good grades. Scholarships seem to be like major league baseball records[17]; there's a unique one available for every participant. The extraordinary cost of education in the United States is, to a small degree, offset by the odd fact that almost everybody at school is paying something less than full price.

The least preferable payment option—again, in my opinion only—is to dip into your savings accounts. Paying out of saved funds is not a bad choice if you only have a few credits left to earn your degree. However, if you're just starting out in your degree program, I would advise against depleting your savings accounts unless it's the last option available to you. Remember that a middle-of-the-road engineering degree can cost in the vicinity of $80,000; if losing that much out of your savings accounts will seriously affect your financial security, then you need to look at other options.

So much, then, for getting in and paying your way. What is the educational experience like for mature-age engineering students? A common worry—I felt it myself—is that when you get in among all these freshmen, you'll feel like a kid who was held down in sixth grade for three years. What is it going to be like sitting in a class full of people half your age?

As it turns out, it's not so bad. In particular, if you're attending night classes, you'll find that a significant proportion of your classmates are also mature-age students who will quite likely have a lot in common with you. Some classes that aren't offered at night may be populated only by the straight-from-high-school

---

[17]   For example, Most Home Runs Hit While Man Wearing Purple Cap Sits in Back Row.

breed of student, but this isn't really as terrible as you might think. You'll probably find that most of the students will simply focus on forming their own social cliques to the exclusion of all else. Remember, these kids are straight out of high school; college was their first major uprooting from home and high school friends; hence, it is much more stressful for them than it is for you, no matter what you might think. You don't need to interact with them very much, but it is helpful to do so—these people will, after all, be doing all the same exams and assignments as you, so it's helpful to build some study-pal relationships if you can.

While we're on the topic of night classes, you should give very careful thought to how much academic workload you can handle. If you have the slightest doubt about your ability to estimate this (particularly if you've been out of school for a long time), I strongly suggest that you start off slowly. For your first semester back at school, just take one course—in a subject area where you're not expecting any nasty surprises—as a confidence-builder. You'll probably be amazed at just how easy these undergraduate studies are for someone who has been in the workforce (particularly the white-collar one) for a few years. The workload of these degree programs is, after all, dimensioned for youths straight out of high school. Despite this, it's important not to dive in and take more credits than you can find time to handle; overstretching yourself will damage your academic record and, more important, discourage you from continuing with the program.

If you have a family, you obviously also need to balance time spent at school with family responsibilities. It's definitely grueling to put in a 40-hour work week, plus school, and then have to come home and handle all the normal duties of family life. Additionally, severe relationship stresses can be created when night classes make your arrival at home late several nights a week. This situation is especially difficult if you choose to attend intensive semesters with a lot of concentrated class time (summer and winter, and other accelerated class options). The best advice I can give you here is to sit down with your spouse and other immediate family before you register for classes, and decide what kind of school schedule your family life can handle.

One final note on being a part-time student: Since it isn't possible for you to complete the degree program at the same rate as a full-time student, I suggest you choose your courses strategically to maximize your ability to learn incrementally. Some course advisers will recommend that you take the "standard" published program and complete the courses on it in the recommended order, half a semester's worth at a time. This is, in my opinion, a dangerous and grossly inefficient way of completing the course requirements. For example, you're going to have at least two semesters of calculus for an engineering degree. Typically, one of these is normally in the first freshman semester and the second one is immediately after it, in the second freshman semester. If you follow the "standard" program at half rate, you are going to end up with an entire semester's dead time in between those two closely related subjects—during which time, you'll forget a lot, thereby impairing your performance in the second course.

Hence, my advice to you is to group together subjects that rely on incremental learning and pursue each set of courses to the end before picking up another thread. In other words, pick two major course work areas (say, mathematics and English) and focus on completing all the required work for those areas of study before proceeding to the next major area. This does mean you'll probably be polishing off some miscellaneous leftover freshman courses like Chemistry 1 and Physics 1 in your last couple of semesters at school, but this is simply your opportunity to be on the dean's list for the last couple of semesters!

Observe, by the by, that there are other reasons besides mere career advancement to catch up on your formal education. If you are working in an engineering capacity without an appropriate degree, you need to be very mindful of professional licensure issues and tread carefully to avoid falling afoul of any legislation. While this doesn't affect embedded engineers as directly as it does, say, architects or civil engineers, every state in the U.S. has specific licensure requirements for the engineering profession. You should check the current laws in your own area; I don't intend to summarize them here, but in general it is prohibited to advertise your services as an "engineer" for hire[18] unless you have the appropriate local license, which in turn requires a formal qualification, some years of work experience, and successful completion of two sets of written examinations. In the United States, the engineering license is called the *Professional Engineer* (PE) license and the requirements, fees, examinations, and other paperwork vary somewhat from jurisdiction to jurisdiction.

---

[18] This is very much a semantic sort of issue. If you advertise yourself as a "programmer," you avoid the letter of the law in every jurisdiction I've studied. What the practical difference might be is a moot point. You might even raise a legitimate argument that you're an artist.

There are, of course, numerous holes that will allow you to work as an engineer without obtaining a PE license. The most obvious of these is that all sins are forgiven (more or less)[19] if you have a real, live PE as a significant partner in your business; they can review, seal, and sign off on documents you provide to the public. There is also an "industry exemption" clause in many areas, which can mean one of two things depending on where you are: either it exempts you from the licensure requirements if you don't offer services to the general public, or it exempts you if you are under the umbrella of an employer who assumes responsibility for what you do. Finally, it is possible to tinker with the wording of your advertising materials in such a way as to avoid specific trigger words and phrases in local legislation.

It's interesting to note that relatively few engineers, *particularly* embedded engineers, are actually licensed PEs. It's equally interesting—and more than a little humorous—to observe that the National Council of Examiners for Engineering and Surveying explicitly acknowledges this fact[20] by stating that "Licensure [...] Sets you apart from others in your profession." I have to admit that I share the viewpoint of many others (including some PEs), viz. that the PE licensing program is a boondoggle—and an expensive one at that.

On a closely related note, you should be mindful of the politics of the situation if you are hired into a large company as an engineer with a nontraditional educational background. I would advise keeping your status to yourself as much as possible; there will be a few people who need to know it, but don't bring the topic up in casual conversation. The credentials issue can easily explode into a big problem if, for example, you perform well and get a good review (and raise or bonus) while one of your traditionally educated colleagues doesn't get much of a reward and consequently becomes disgruntled. You should also be prepared, if necessary, to accept a job title that isn't the one you applied for. (For example, you may be hired into a research engineering position, but for political reasons your official title might be "electronics programmer.")

---

[19]   Again, please don't take my statement here as gospel. This is emphatically *not* legal advice; it is merely a warning to let you know that there are complications ahead. You should research local laws before embarking on any course of action.

[20]   As of the time of writing, this quotation could be found at *<http://www.ncees.org/licensure/licensure_ for_engineers/>*.

In summary of all the previous text: It's possible to avoid most or all of the state's paperwork requirements if you either play word games, or you live inside the shield of a person or entity who can keep you out of harm's way with respect to licensing. Nondegreed engineers, however, simply won't be welcome at the country club. Since anyone who is competent to do an engineer's job will most likely have very little difficulty with the academic requirements of a bachelor's degree, it's well worth the effort to go back to school and fill in your background if you have the opportunity. If your employer will pay tuition, you have no reasonable excuse not to go back to school. Becoming a PE is optional, but might increase your marketability in certain respects.

2.3   I Write Software—How Much Electronics Must I Learn?

Before we embark on this topic, it's time for a short humor break illustrating the potential silliness of embedded software development.

The 999,999th Monkey Theorem

Emile Borel's 1909 book on probability stated that a monkey hitting random keys on a typewriter would eventually type every book in France's national library. You're probably much more familiar with an anglicized restatement of the theorem, commonly put this way: "A million monkeys at a million typewriters will eventually produce the complete works of Shakespeare." (There are numerous other ways of saying the same thing, of course, and many people prefer to assign infinite monkeys to the task—however, this presents us with the intractable Infinite Banana Paradox, so we will consider only finite monkeys.)

A frightening amount of software is clearly generated using the million-monkey approach, which is sufficient evidence to account for the fact that almost every software release since the first set of cards for the Jacquard loom has been delivered late. (We're sorry, Sweater 1.1 has been delayed due to problems with the cable stitch subroutine.) Now, the difficult part about the monkey method is not setting the monkeys to work, but filtering the results. Scanning random text for Shakespeare is relatively easy, because the desired output is well-defined; you simply compare the output against the known works of Shakespeare and there you have the answer.

Scanning the output of a million monkeys for software that meets the design specification is a much harder task, because you don't know exactly what you're looking for.

The 999,999th Monkey Theorem goes like this: Given a desired output and a functioning monkey engine, there will be a finite set of "correct" answers (i.e., answers that satisfy all the constraints for "perfect" output—"To be or not to be, that is the question."). There will also be an infinite set of answers that are useless gibberish ("Thabahq892a qw[t980q324[!"). However, for any correct answer, there are an infinite number of possible variants of that answer, all of which are SLIGHTLY incorrect but SUPERFICIALLY appear to be correct ("To be or not to be that, is the question."). Since the infinite number is infinitely larger than any finite number, there is a negligible probability that any answer that merely appears to be correct actually will be. Unfortunately, this also means that the discriminator logic on our monkey-driven code engine can only, at best, decide "gibberish" versus "possible solution"—it can't pick the correct solutions out of the set of possibles, and the odds are against it happening onto a perfect answer.

Therefore, we conclude that any functional software generated using the million-monkey approach is practically certain not to be working correctly, and we can say this with confidence even if we have no idea how it's supposed to work or what it actually does.

A corollary of this is that the only form of transportation I trust is a horse, which was admittedly designed through random processes, but has had a much longer beta test period than any man-made vehicle. The real worry is, of course, that the rewards for being the 999,999th monkey regularly are much larger than the rewards for being the millionth monkey once.

For sample output from a real monkey cluster (albeit with somewhat fewer than a million nodes), visit the following URL: <*http://www.vivaria.net/experiments/notes/ documentation/*>.

Among many fiercely argued religious questions is the issue of how much electronics knowledge an embedded engineer needs to have in order to work effectively. Some people maintain that for pure firmware engineers, no hardware knowledge is necessary beyond a simple understanding of register-level system behavior. At the other end of the continuum, some people argue that they would never hire an engineer, even for a wholly software role, unless he or she has a good practical grasp of both analog and digital design.

The real answer to this question is, of course, somewhere in the middle—and highly dependent on what sorts of projects you intend to do. To a certain extent, the size of the company you're aiming to work for is also a factor; if you're working for a small company or in a small team within a bigger company, you'll be under increased pressure to be self-reliant and able to solve problems that are "across the divide" from your job description (for more on this topic, see Chapter 6). You don't need to be a guru, but you do need to be able to understand the behavior of the other parts of your system so that you can predict how the system will behave given a certain stimulus from your software. As you get down close to the bare metal (writing device drivers, for instance), it can be practically impossible to develop relevant debugging skills unless you're able to hook up a scope or logic analyzer and understand what the actual hardware is doing in response to your code.

While I'm on this point, let me take a moment to state most vehemently that the idea of either software that's independent of hardware or hardware that's independent of software is dangerous and silly. Neither component can operate in a vacuum; you can design a fabulously complex and powerful piece of embedded hardware, but until it becomes a platform supporting an operational piece of software, it's just a piece of laboratory junk. Likewise, you can develop embedded software in some non-native simulation environment, but until it's loaded onto real hardware, it's a computer science project, *not* a product. (This does not altogether disparage simulation, by the way, but the premise of software simulation is that the *simulation* environment emulates salient features of the actual hardware on which the final product is expected to run.) As a design engineer in a large company, it irks me no end to see the hardware team build something "perfect" and toss it over the wall to a software group. Of course, the software group has developed "perfect" software that runs just fine on the emulator and maybe even the first prototype PCB. Naturally, when the perfect software meets the perfect hardware, unforeseen problems arise.

So what, exactly, should you know? Here is a short list of the skills and knowledge that I would consider essential but which might not be obvious to a newcomer. Observe that not all of this is explicitly included in a BSEE degree.

- At least a rudimentary understanding of how to route power planes and the consequences of poor layout on power quality. Even if you never personally lay out a board, you need to be able to debug problems caused by PCB layout snafus. Your BSEE courses will most likely not teach you anything terribly practical about PCB layout techniques. A very useful book—with much more information than you're ever likely to need—is *High-Speed Digital Design: A Handbook of Black Magic* by Howard Johnson (Prentice Hall PTR, April 1993, ISBN 0-1339-5724-1).

- In the same vein, you should have a basic understanding of how PCB routing can affect signal propagation.

- The ability to read a schematic.

- An introductory-level understanding of the DC characteristics of diodes, bipolar transistors, FETs, op-amps, and comparators.

- Comprehension of the different types of I/O configurations on digital parts such as microcontrollers; open-source, open-drain, full totem-pole, the presence or absence of protection diodes, and so forth. You particularly need to understand issues related to level shifting (today's systems often have mixed I/O voltages) and driving different sorts of loads. I've seen far too many systems that (for example) drive inductive loads like relays with no attention to the nature of the load, leading to all sorts of bizarre occurrences as relays open and close.

- Some simple techniques for mitigating ESD susceptibility; placement of spark gaps, series resistors and capacitors to ground where appropriate.

- The ability to operate a SPICE simulator is frequently helpful, but not absolutely essential.

- Practical hardware debugging skills (see Section 2.5).

Having said all that, if you have absolutely no electronics knowledge at all, and no desire to acquire any, then I recommend you set your sights for embedded jobs high in the software abstraction hierarchy. In terms of this book, that means you should probably skip to Chapter 4. Such systems do, of course, require the involvement of people who are knowledgeable in both software and hardware; however, the scale of these systems is so large that the hardware-savvy people are likely to be concentrated in the operating system and device driver development layers. In such enormous projects, there is usually a place for people working in the application layer who never need to know anything about the hardware except how to use the abstraction APIs supplied by the operating system team.

2.4 Educational Traps, Dead-Ends, and Scams to Avoid

When I read my morning's junk mail and see the usual plethora of offers for fake or meaningless degrees, the grandiose claims and untrammeled avarice of the authors always remind me irresistibly of Charles Dickens' tirade against Yorkshire schools of his age.

Of the monstrous neglect of education in England, and the disregard of it by the State as a means of forming good or bad citizens, and miserable or happy men, private schools long afforded a notable example. Although any man who had proved his unfitness for any other occupation in life, was free, without examination or qualification, to open a school anywhere; although preparation for the functions he undertook, was required in the surgeon who assisted to bring a boy into the world, or might one day assist, perhaps, to send him out of it; in the chemist, the attorney, the butcher, the baker, the candlestick maker; the whole round of crafts and trades, the schoolmaster excepted; and although schoolmasters, as a race, were the blockheads and impostors who might naturally be expected to spring from such a state of things, and to flourish in it; these Yorkshire schoolmasters were the lowest and most rotten round in the whole ladder. Traders in the avarice, indifference, or imbecility of parents, and the helplessness of children; ignorant, sordid, brutal men, to whom few considerate persons would have entrusted the board and lodging of a horse or a dog; they formed the worthy cornerstone of a structure, which, for absurdity and a magnificent high-minded laissez-aller neglect, has rarely been exceeded in the world.

—Charles Dickens, from the preface to *Nicholas Nickleby*

EDUCATION—At Mr Wackford Squeers's Academy, Dotheboys Hall, at the delightful village of Dotheboys, near Greta Bridge in Yorkshire, Youth are boarded, clothed, booked, furnished with pocket-money, provided with all necessaries, instructed in all languages living and dead, mathematics, orthography, geometry, astronomy, trigonometry, the use of the globes, algebra, single stick (if required), writing, arithmetic, fortification, and every other branch of classical literature. Terms, twenty guineas per annum. No extras, no vacations, and diet unparalleled.

—Charles Dickens, *Nicholas Nickleby*

In Section 2.1, I already made passing reference to the BSEET program and its inadvisability for people considering a long-term career in embedded engineering. However, there are numerous other bear traps lurking for the unwary student. These traps range from courses that simply aren't as useful as they appear, to so-called life experience degree programs that are downright fraud (more on this later). The problem of choosing a degree program in the United States is exceptionally difficult to unravel because the tertiary education system in this country operates on an apparently unique free-market principle with no central government regulation. Practically anybody can collect fees and issue "degrees." A third party (say, a prospective employer) can theoretically judge the value of a degree by whether the issuing institution is accredited by someone reputable. However, this becomes something of a chicken and egg situation, because any fraudster who can open a degree-issuing "school" can also create a private accrediting agency to "accredit" that school.[21] So, to evaluate the worth of a degree, you really need to know both by whom the issuing institution is accredited and the bona fides of the accrediting authority.

Because of this, combined with the plethora of different curricula and performance claims made by various institutions, it can be excruciatingly difficult to evaluate the relative merits of different higher education programs. Obviously, an Ivy League degree is (usually) easy to distinguish from a "print your own diploma, pay only for paper!" sort of operation, but in between those extremes there is a very broad spectrum of cost and utility. Let's not forget, also, that once you finish your degree, you have to be able to convince an employer or another college of the utility of this qualification.

As a general heuristic, I'd say that the closer you can approximate a "normal" educational program (simulating the progression of courses and events that would

---

[21] An interesting article on this topic, with some historical background, can be found at *<http://www.degree.net/guides/accreditation_guide.html>*.

have occurred if you had gone into this program immediately after high school), the easier your life will be. College bureaucracies are relatively ill-equipped to deal with out-of-the-box situations, especially at the undergraduate level. If you need to transfer credits or—worse still—get your degree recognized in a different country or evaluated by a licensing authority or employer, you will have a much smoother time if your paperwork is simple to understand.

Many people contemplating a career change turn to distance learning programs by mail or Internet so that they can more easily fit school into a busy work and home schedule. The first problem to consider, then, is how to evaluate distance learning programs. Now, what I'm about to say here is a rather unfair generalization, but in general I would advise extreme suspicion and caution when looking for an electrical engineering degree that's offered entirely by distance education. Engineering is a practical discipline—lab work is an important part of the degree, and this is difficult to do by mail or Internet.

If we exclude "colleges" that are generally known to be degree mills,[22] there are very few institutions indeed that offer a completely online electrical engineering degree program, and you could spend a long time trying to find them. Engineering, as a practical discipline, has a laboratory component that is very difficult to provide by way of distance education. (Note, by the way, that doing a legitimate online program of this sort will require the assistance of an engineer or other qualified professional—perhaps your manager—local to you. This person will be the proctor who oversees your exams and other graded, timed work. Different schools set different criteria for what constitutes an acceptable proctor.) If you're evaluating such programs, make sure you check that the program specifies that it is ABET accredited, and cross-check this by visiting ABET's site and searching for the institution. ABET schools should all be regionally accredited by one of the six major regional accrediting agencies that operate across the United States,[23] so you don't need to worry about that aspect of school legitimacy.

---

[22] An enjoyable read from respected authors on this topic is *Degree Mills: The Billion-Dollar Industry That Has Sold Over a Million Fake Diplomas* by Allen Ezell and John Bear (Prometheus Books, January 2005, ISBN 1-5910-2238-X).

[23] The Department of Education's Office of Postsecondary Education maintains a handy searchable database of accredited institutions (read their definition to understand what this means in context) at *<http://www.ope.ed.gov/accreditation/>*.

Be aware that there are several institutions—I can't say authoritatively whether they are generally recognized as degree mills or not—that have invented engineering "accreditations" for themselves. One of these institutions, at least until quite recently, advertised its "National Siciety (*sic*) of Accredited Engineers" laurels on its website. Other institutions make vague statements, such as "licensed by the State of Wyoming, about licensure or accreditation."[24] Such a license, if it even exists, simply refers to a license to operate as a secondary or post-secondary school; it has nothing to do with regional accreditation for the school and *absolutely* nothing to do with the one accreditation, which is ABET, you're really interested in.

Given all this complexity, you might be sorely tempted to wonder whether it really matters which college you choose to attend. After all, most employers probably won't even check to see whether your alma mater exists, let alone verify whether it's regionally accredited. There are certainly thousands of people in the workforce with meaningless diploma-mill qualifications. Unfortunately for them—and for you, if you take this path—this is rather like sitting on a hand-grenade with the pin pulled. All it takes is one disgruntled work colleague asking questions, and you could be fired for cause. It's definitely not worth the risk, and it's not even necessarily cost-effective—many diploma mills charge the same sort of fees that you would pay at a reputable school.[25]

One well-respected reference on the topic of distance learning is *Bear's Guide to Earning Degrees by Distance Learning* by Mariah P. Bear and Thomas Nixon (Ten Speed Press, January 2006, ISBN 1-5800-8653-5). This book is updated regularly (currently in its sixteenth edition) and should be considered an essential starting point if you are considering distance learning. As a companion to their books, Bear et al. also run an informative website at <*http://www.degree.net/*>.

Now, this should be self-evident, but never fall into the trap of believing that any educational offer you receive by email is in any way above suspicion; in fact,

---

[24] I use Wyoming as my example advisedly since this state is notorious for having lax standards for issuing licenses to schools. You can read the relevant legislation at <*http://legisweb.state.wy.us/statutes/ titles/title21/c02a04.htm*> if you're interested. However, the whole legislative issue is a whack-a-mole game at best—once you tighten the rules in one jurisdiction, the scam operators just jump across the border to the next least stringent area.

[25] I must shamefacedly confess that when I started to look at distance education programs, I didn't understand the U.S. accreditation system, so I didn't ask the right questions of the right people. As a result, I wasted substantial sums on an institution that is, while not classified as a diploma mill, definitely a borderline case. You won't find it on my resume.

I would take the opposite stance and assume that any offer thus received is bogus. All the job-hunting websites (and other sites, for that matter) in the known universe will sell your contact details to anybody willing to pay for them. They will also sell "targeted" email campaigns to anyone with the wherewithal to finance the operation. As a result, if you have ever posted a résumé or applied for a job online (and even if you haven't), you will certainly receive regular junk email touting "life experience degrees." While most colleges will in fact grant some undergraduate credits for life (read: *work*) experience, no legitimate institution will give you a diploma purely on the basis of experience. Getting credits for life experience at a real college can be quite a chore; the requirements range from simply obtaining a letter from your manager stating what it is you do in your day job, to writing a 5,000- to 10,000-word essay on the subject matter in question and/or possibly taking an examination.

Finally, there is a range of IT qualifications, such as MCSE, CNE, CNA and so forth, which are advertised quite heavily and frequently offered as internal training options inside a corporate environment. All too often, people who express an interest in migrating into "programming" or "engineering" jobs will receive a recommendation to take one of these courses. This is quite mystifying to me; I can only conclude that in the mind of a nonengineer, all job titles containing the word "engineer" are inseparably conflated. Please don't waste your time or money on such studies; they are strictly industrial qualifications that pertain to IT jobs (a "network engineer" is not an engineer in the sense that we mean it in this book, and is certainly not an embedded engineer by any stretch of the phrase). Worse, most of these certifications need to be renewed annually, at some considerable cost. As a side note, I'd also like to point out that the sorts of jobs to which such qualifications pertain are *precisely* the category of high-tech positions that have been migrating en masse to offshore support centers in recent years.

So much for the things you definitely *don't* want to study. On the other hand, there are skills available from formal education that engineering students appear to regard as irrelevant but are actually of great importance. The next section discusses some of them in detail.

2.5 Practical Skills You'll Want to Acquire

Regardless of how far you eventually intend to specialize, there is a small core of baseline skills that will benefit you greatly in any sector of embedded engineering. Some of these skills are taught explicitly in college, some of them are mentioned peripherally but not examined in any great detail, and the remainder are acquired and honed exclusively through practical experience.

I hope the fact that you're reading this book—and presumably somebody paid to buy it—helps to demonstrate that one of the most important skills you can acquire is to learn the lost art of effective reading and writing in the apparently dead language known as English. Most engineers will probably not author full-length books, but any good engineer will write many thousands of words of technical documents in their careers, including the following:

- Product specifications, explaining to marketing and the people who write your product manuals exactly what the product will do

- Protocol specifications, explaining to other engineers how to talk to your product

- White papers, describing to other engineers what you've been working on and useful discoveries you've made

- Patent disclosures

- Instructions to subordinates

- Articles for technical journals (Being published in this way can *substantially* improve your visibility within an organization—think raises and promotions.)

- Debugging information, communicating with vendors, quality assurance technicians and engineering colleagues to solve complex problems

- Justifications (dare I say it) for taking a specific course of action (When things get out of control and the recriminations start to fly, the one with the best paper trail usually wins.)

Most engineering students—in fact, I can generalize wildly and say most science students—regard language skills as a mere waste of time, required by an idiotic college bureaucracy. Unfortunately for these students, it has been my experience—very rarely contradicted[26]—that engineers who can write intelligible and concise documents are precisely the same engineers whose specifications are complete and easy to understand, whose code is well-structured and simple to read, and who have little difficulty communicating effectively with co-workers.

You'll note that the BSEE breakdown I gave in Section 2.1 only shows six credits of English. This is more or less representative of the average BSEE degree; you can add a little more writing and speaking experience in the liberal arts electives, but no matter how you wiggle your course schedule, a bachelor's degree in engineering is only going to give you the barest possible taste of formally taught language skills. *You need to practice this.* Voracious reading—not just technical books, but fiction and nonfiction by good authors—is essential; reserve time for it in your week (hints: the bath is a great place to relax and read, and Project Gutenberg has numerous free electronic texts you can download to your PDA and read anywhere). Practice writing wherever you can; before I was ever published, I honed my skills largely in banter—technical and otherwise—on Usenet and, previously, Fidonet.[27] Good language skills can float a competent engineer significantly above their colleagues.

Another essential skill set on which you will not touch significantly in formal education is PCB layout and an understanding of DFM (design for manufacturing) concepts. Both of these are specialties in their own right, but from the perspective of the embedded engineer, they're closely related and can be learned at the same time. Although a design engineer in a company of even modest size probably won't spend any significant time working on layouts, there's a great deal of practical value in understanding at least the basics of what is involved in the processes of designing, laying out, and populating (stuffing) PCBs.

---

[26] I'm not counting people for whom English is a second language in this statement. Wherever I write "English" in this section, you can substitute "the engineer's native language." It's language skill, not specifically English skill, that I'm talking about. However, if only because the majority of scientific and technical publications are in English, it would be prudent to study this language if you don't speak it natively.

[27] A pre-Internet worldwide network of bulletin board systems; my system was ZWSBBS, 3:634/396. Fidonet still exists (though greatly shrunken since my days there) and is connected to the Internet at various points—see *<http://www.fidonet.org/>* for more information.

If you work at a large company, you'll have opportunities to talk to the manufacturing engineers and PCB layout artists and glean a lot of useful wisdom; you should seize these opportunities when available. You don't necessarily need to annoy people with incessant questions; merely listening carefully at design reviews will greatly improve your understanding of the issues. There's a very large body of succulent information tidbits that seem intuitively obvious once you hear them, but which you might never think of yourself; for example, the need to keep ceramic surface-mount capacitors away from the edges of PCBs to avoid cracking them during depanelization; ensuring that you leave sufficient clearance around a surface-mounted microcontroller to allow use of a test clip; understanding how components drift and/or self-align on their pads when they go through the infrared reflow oven; and so on.

You'll also need a lot of practical software debugging experience. A good starting point for this is simply your college courses—if you work at the computer science and other programming-related course work with sufficient assiduity, you'll have a reasonable start on the methodology of software debugging by the time you graduate. However, one issue for which school will leave you almost completely unprepared is how to jump in at the deep end and start maintaining someone else's large code project.[28]

It is practically guaranteed that your first job (and every subsequent job, for that matter) is going to involve maintaining some legacy code. You can be very lucky and find code that is accurately documented and well-commented, or you can encounter code (as I have) that contains no comments at all and is structured in such a complex and bizarre fashion that adding or removing a single comment line can cause the compiler to abort with an internal error. The most useful learning technique I have found for this is to take an open-source project and add specific simple functionality to it. For instance, you might set yourself the goal of modifying the IDE driver in the Linux kernel so that it turns on a

---

[28] A general disadvantage of learning in the school context is that you know that the problem you've been assigned has a solution, you know that you have given the skills and the time to find that solution, and you can make inferences (from where your class is up to in the curriculum) as to which specific skills are likely to be relevant to the problem. Real-world problems don't have any of these guarantees or hints; you're on your own with a clock (and budget) steadily ticking away and management breathing down your neck. Enjoy your time at school; these are the easiest problems you will ever be asked to solve.

red LED whenever a write operation is requested from the application layer, or a green LED for read operations. A more challenging project might be to take an open-source operating system, such as NetBSD or eCos, and port it to a new board. In any case, the goal is to understand enough of the existing code to know where to insert your changes and modify the code as nonintrusively as possible to avoid creating bugs.

Finally, most embedded engineers are going to need at least rudimentary laboratory skills. This includes being able to solder up prototypes and operate an oscilloscope, signal generator, and spectrum analyzer. The introduction you'll receive in college is probably adequate grounding in this; you'll gain much greater facility with experience, of course. In this day and age, it's also very useful to be able to work with surface-mount components. You can practice this very inexpensively on any junked piece of hardware—an old PC motherboard, a DVD player, or any similar board with lots of parts on it.

I'm sure you've noted the recurring theme here: even if you learn some skills theoretically at school, I'm exhorting you to find opportunities to keep working at example problems requiring those skills in real life. Furthermore, a lot of what you need to know can only be acquired through dealing with actual problems. Engineering is a *practical* discipline. Steep yourself in homebrew projects to keep your skills fresh and ready for action; you're also demonstrating self-motivation to future employers.

3

Teaching Yourself, Bottom-Up (Small Embedded Systems)

3.1   Target Audience

One of the most frequently asked "newbie" questions asked in embedded engineering forums is (paraphrased), "Which microcontroller should I learn about in order to have directly marketable embedded skills?" This question is fairly commonly posed by technicians and engineers with circuit design experience, who are looking to extend their skill set upward into firmware. This chapter will provide such people with some idea why the newbie question can't be answered in any simple way, and some useful recommendations as to which platforms you might choose for experimentation and learning purposes.

This is an appropriate moment to make a very important point to those people who work primarily in hardware design: Not only is it practically mandatory for a latter-day hardware engineer to understand, at least partially, the software layer above their head, but having a good understanding of what can be achieved in firmware will make your circuit designs more efficient. There are problems that are much easier to solve in firmware than in hardware, and a good design engineer—software or hardware—will keep this in mind so that the overall design can be tuned with respect to the features that will be implemented in software versus those that are dealt with in hardware.

To take a trivial example—most microcontrollers that implement a hardware UART expect to be connected to an off-the-shelf RS-232-compatible line driver/receiver chip (the MAX232 is a common candidate, and it's available from several sources). If you're connecting a microcontroller to something else on-board via the UART, and you don't intend to shift the signal to RS-232 levels, the software will be simpler and much more cycle-efficient if you design the other end of the communications link to accept inverted input.

In any case, the short answer to the "Which micro should I learn for fortune and glory?" question is that learning how to work with any one particular micro-controller won't lead you directly to a job.[1] When you come to a new job, you're going to find a certain body of legacy technology implemented on some micro or other (probably some hoary old device you would never choose in your wild-est nightmare[2]), and perhaps a newer generation based on a different family of micros—and when you start working on your own completely new designs, you may choose a different family yet again. All sorts of factors will affect your choice of micro; future availability, pricing, contract terms with distributors, power sup-ply limitations, performance requirements, peripheral requirements . . . and so the list goes on.

You might occasionally see people asking if there is any reason to learn how to develop on 8-bit cores, since the 32-bit cores are getting cheaper every day. These people don't take account of the fact that for high-volume consumer appli-cations like VCRs and CD players (which don't benefit at all from a high-end microcontroller), a few pennies in cost translates to massive annualized savings. These applications will not migrate from 8-bit to 32-bit cores until there is no cost difference at all.

Note that this is not quite the same thing as saying "never," and it's worth explaining some of the issues so that you can understand the factors at work here. As die sizes shrink, the packaging and bonding costs (bonding wires from the silicon to pins that go out to the real world) start to dominate the cost of the device. Raw 8-bit dice will always be cheaper than raw 32-bit dice due to their lower transistor count, but by the time the chips are packaged for sale, the price difference may be almost invisible. For some applications, it is already foreseeable

---

[1]   Sometimes, a particular job will "require" experience with a specific microcontroller. Job descrip-tions constrained in this way indicate a lack of understanding on the part of the job poster. For any competent engineer, learning a new microcontroller is a trivial exercise; learning the *system* is the time-consuming part, and this training time can't be reduced significantly by mere familiarity with a few of the component parts.

[2]   This has been the case with every day job and contract job I have ever taken. Existing code and hardware are, to the new hire, like arranged marriages—they might work out for the best, but trepi-dation and caution are the prudent approach.

that there will be no real cost benefit to be realized by staying with an 8-bit architecture. There is a complicated and rather interesting set of simultaneous equations at work here. Individual chip dice (raw unpackaged chips sliced off the wafer) are priced mostly according to the number of processing steps in the "recipe" and the number of good dice yielded by each wafer. As chip geometries get smaller, the number of dice per wafer goes up. Additionally, the *absolute* cost differential between an 8-bit core and a physically larger 32-bit core goes down, since the *absolute* size of each is shrinking. (The relative size difference is still the same; if the 32-bit core was 30% larger before the shrink, it will still be 30% larger afterwards. However, since each feature costs less after a die-shrink, the financial impact of that size difference is reduced.) On the flip side, however, smaller geometries can lead to reduced yield, since a single defect in the wafer is likely to affect a larger number of dice; this factor puts upward pressure on the cost of small-geometry chips. The net result of all this banter is that as manufacturing processes improve, the cost of a core—of either flavor—is asymptotically approaching zero. As both 8-bit and 32-bit cores get closer to that magic spot, the cost of the core is starting to disappear behind other factors.

For parts with a lot of on-chip memory, for example, the core cost is more or less irrelevant compared with the cost of the memory array. However, this question becomes more complicated still because the processes used to make memory are not the same as the processes used to make microprocessor cores. A microprocessor requires several layers of interconnects due to its complexity—it needs to route large buses around multiple peripherals, and in many cases needs additional special handling to meet mixed-signal requirements. A memory array, on the other hand, is simply a two-dimensional matrix; it doesn't need so many layers. Since all areas of a chip have to go through every process required by any single part of the chip, this means it is much more expensive (dollar per kilobyte) to put memory on the same die as the processor. At a certain size cut-off, therefore, you will find that manufacturers switch to a stacked-die configuration—the memory array is a separate chip, and the microcontroller is mounted on top of it.[3] Bond wires connect the two.

---

[3]   For any chip where this issue is a consideration, the memory section will be geographically much larger than the microcontroller core.

The principal downsides here are (a) the cost of bonding the two parts, (b) vulnerability to vibration damage, (c) reduced speed, and (d) increased RF emissions. It is very instructive, when comparing the prices of different (large) micros with similar features, to drill down into the details like this and determine exactly what disdavantages you're going to experience from choosing the cheaper part. Some applications might not be affected, but anything with tight RF requirements (a GPS receiver or cellular phone, for instance) needs to think very carefully about the cost/benefit equation of choosing stacked-die parts.

Leaving this interesting digression to one side, if you've got a solid practical electronics background—and I'm speaking here particularly to hardware design engineers, but also to technicians—you've got a big advantage over software-only people in the sorts of applications where 8-bit processors are commonly used. The reason for this is that you're already used to extracting state information out of a complex system using hardware tools that typically only allow you to look at the system through a tiny window. You've also gained a comfortable level of familiarity with what can go wrong with the system on the outside, from a voltage/time perspective.

Hence, if you can drive an oscilloscope or logic analyzer competently, you've already acquired one of the most important skills you're going to use in firmware development: being able to infer a great deal about the system being debugged by peering through a relatively small aperture. Although it's nice to have a full symbolic debugger, this luxury is often not available. Sometimes it's too expensive to justify purchasing the necessary hardware and software, and sometimes you simply can't reproduce the problem(s) in your code when it's running on an emulator.

In the remainder of this chapter, I'm going to introduce you to a few popular microcontroller cores, and describe some of their features, both enjoyable and execrable. This chapter is not, of course, an up-and-running guide to using any of these parts. It is a high-level overview of what's inside these parts, where you might use some of these device families, and where their design strengths lie. Where possible, I'll also provide you with pointers to places where you can obtain hardware and software to do your own development and experimentation.

On a related note: In this chapter, I mention several specific vendors and products by name, and I quote some approximate prices. I'm not endorsing these vendors in the same sense that an athlete might endorse a sports shoe company

("Buy Fred's Sneakers! Their check to me cashed really fast!"). I am explicitly not exhorting you to buy anything in particular; I'm merely providing these specifics as a convenience feature so that you can quickly find and read about the products that I've used and found well-suited to unprogrammed learning on a budget.

3.2　Intel® (Et al.) 8051 Variants

Intel's 8051 architecture is an industry-standard platform referenced in almost every engineering course since the beginning of the microprocessor era. When spoken of in polite company (and even in books like this one), you'll generally see the number 8051 prefixed with the adjective "venerable" or a synonym. Personally, I would much rather describe it as "decrepit," but it's a solid fact that the 8051 is still the world's bestselling 8-bit microprocessor core. Barring some major religious uprising, this is likely to remain true for as long as the world is still manufacturing and using 8-bit micros.

I'm including a discussion of the 8051 in this book for two main reasons and one minor reason:

1. It pops up in all sorts of apparently unrelated applications—many application-specific standard products (ASSPs), for instance, have an 8051 core. The chance that you'll run across an 8051 variant in your career is thus very good indeed. Being familiar with the core's capabilities, if nothing else, will help you decide how to put together your design.

2. If you can work efficiently with the 8051, you can work with anything, so it's not a terrible architecture to learn on.

3. (Minor reason)—The 8051 happens to be very efficient at tasks that involve bit-level data manipulation. Clock for clock, it is faster at these operations than many 32-bit microcontrollers. You may find this fact useful, since the 8051 is also very cheap.

For educational purposes, the principal advantage of the 8051 family is its ubiquity. From the point of view of real, commercial projects, the principal stated advantage of the 8051 is its availability from multiple sources. It is widely claimed that this makes the 8051 a truly multisourced part, and that you have something approximating a guarantee that you will have a migration path to another vendor if a particular source is giving you problems.

This conventional wisdom is, for the most part, *not* strictly true, and it irritates me somewhat when people make this comment without sufficient qualification. While it is certainly true that a few exceedingly vanilla standard parts are available in pretty nearly identical forms from several vendors, all of those vendors also offer more specialized parts. Some of them feature optimized microcode with fewer machine cycles per instruction; some have higher maximum clock speeds, more RAM, on-chip debugging capabilities, extra UARTs, and so on. Once you start looking at ASSPs[4] like digital camera ICs, USB keyboard and mouse controllers, USB-to-Flash-media interface chips and so on, there is no standardization at all beyond the bare instruction set; certainly, there are no drop-in replacements from different vendors. As a result, the 8051 is really only a true multisource part (in the sense of drop-in pin-for-pin replacement) if you take great care to avoid using any vendor-specific special features—in which case, why should you pay for them?

For the purpose of giving you a quick introduction to the 8051 family, I'll describe the basic Intel 80C51 automotive part.[5] This microcontroller is available in either 40-pin DIP or 44-pin PLCC packaging (the pinout of both is industry-standardized). It offers 4 KB of on-chip EPROM or ROM, 128 bytes of RAM, two 16-bit timer/counters, five interrupt sources, one UART, and a clock speed of up to 16 MHz. For development purposes, you would normally use either an ICE, or the 87C51, which is the EPROM variant. (Note that 8051 part numbering, for vanilla parts anyway, works similarly to nonvolatile memory parts—as a rule, 87xxx parts are EPROM, 89xxx parts are Flash.) Figure 3.1 shows the pinout of the 40-pin DIP 87C51; numerous other variants are identical, or almost so.

---

[4]   Application Specific Standard Product—for example, digital camera ICs, USB memory stick drive ICs, CompactFlash controllers, infrared remote control chips, dedicated CD-ROM servo control chips, printer controllers, and so forth.

[5]   Information in this section is redacted principally from various reference circuit designs, as well as "80C31BH/80C51BH/87C51 MCS® 51 CHMOS SINGLE-CHIP 8-BIT MICROCON-TROLLER" (Intel, 2004), and "MCS® 51 MICROCONTROLLER FAMILY USER'S MANUAL" (Intel, 1994).

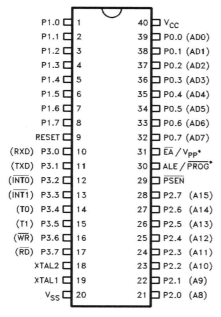

Figure 3.1 40-DIP 8051 pinout.

The pin count is dominated, as you see, by the general-purpose I/O (GPIO) ports. There are four 8-bit ports in the standard 8051, providing up to 32 available I/O lines. Ports 0, 2 and 3 are, however, multiplexed with other functions, shown in parentheses. (Note that there are still other alternative uses of some pins when the part is being programmed in a suitable burner; those uses are not shown here. The baseline 8051 series parts are not in-system programmable.)

Be aware that not all of these GPIOs behave identically, even when their associated multiplexed peripherals are not enabled! It is a rookie mistake in an 8051 design to assume that the the drive characteristics of all GPIOs are identical. In particular, you must note that port 0, when being used as a general-purpose port, has open-drain outputs with no internal pullup resistors. (The behavior of port 0 is different when external memory is being accessed; see the following for more details on that.) Ports 1, 2, and 3 all have on-chip pullups; if you are using them as inputs, and an external device pulls the pin low, the 87C51 will source current.

The following are other miscellaneous pin functions.

- *RXD, TXD* – Receive and transmit data for the on-chip UART, respectively.

- *_INT0, _INT1* – External interrupt request lines.

- *T0, T1* – External clock signal inputs for timers 0 and 1.

- *_RD, _WR* – Read and write strobes, respectively, for off-chip RAM.

- *_PSEN* (Program Store ENable) – Read strobe for off-chip program memory.

- *ALE/_PROG* – Address Load Enable pin, used by multiplexed address/data bus when accessing external memory. Also used when EPROM parts are being programmed in a specialized burner.

- *RESET* – Resets the microcontroller (note that it is active high, which is slightly unusual).

- *_EA/Vpp* – If this pin is strapped low, the micro will boot off external program memory instead of on-chip memory. This is a supremely useful feature if you've come across a boxload of used OTP or ROM 8051s; just tie this low and use the part in ROMless mode. I've come across commercial appliances—particularly cheap modems, and certain aftermarket automotive accessories—that used all sorts of recycled scrap/surplus 8051 parts in this way. If you intend to let the device run off internal memory, you should strap this pin high. Note that some 8051 variants may disable this function partly or entirely if code read-out protection is enabled.

- *XTAL1, XTAL2* – Crystal oscillator amplifier input and output.

In addition to the features previously described, the 8051 has a multiplexed address/data bus allowing the device to access up to 64K of external static RAM (SRAM) and 64K of external program memory. These two memory types live in different address spaces, since the 8051 is a Harvard-architecture device.[6]

The method by which each of these external memories is accessed is basically similar but slightly different in the details. First, here's the process by which the

---

[6]   In case you don't remember Computer Science 101, in brief: von Neumann architectures put code and data in a single address space, and Harvard architectures put code and data in separate address spaces. This has all sorts of implications to the embedded programmer—particularly to the inexperienced embedded programmer using a high-level language that hides some of this detail.

8051 accesses external program memory, either using the MOVC instruction (more on this later) or by jumping to a location in off-chip memory.

- The micro brings ALE and _PSEN high.

- The low 8 bits of the desired fetch address are output on port 0. In this mode, the port uses strong internal pullups; you don't need external pullups if you are only using port 0 as the address/data bus.

- ALE is brought low.

- _PSEN is brought low and simultaneously, Port 0 switches to input mode to read the instruction. (Port 2 is still outputting the high byte of the desired address.)

In a typical 8051 circuit employing external memory, ALE is routed to the clock input of an octal flip-flop such as a 74HC373, and P0.0..P0.7 are routed to the D0..D7 inputs of that chip. The Q0..Q7 outputs of the 74HC373 form the low 8 bits of the memory address bus (A0..A7); the high 8 bits (A8..A15) come directly from P2.0..P2.7 on the 8051.

The A0-A15 lines are routed directly to the address inputs of your memory or memories. The _PSEN signal goes to the _OE (Output Enable) line of your EPROM or Flash memory chip, and _RD and _WR should be run to the _OE and _WE (Write Enable) lines, respectively, on your SRAM. Separate chip selects (if any) on the memory devices should be tied to their active states. All of this sounds rather complex, and looks quite impressive if you draw it on paper, but if you look closely you'll see that in a large number of cases you can implement external RAM and ROM with no additional components besides the 74HC373 latch. In fact, some manufacturers even sell SRAM and/or EPROM chips (and exotic combined SRAM/EPROM or SRAM/Flash devices, in some cases!) that already have the address demultiplexer latch built in, so that 8051 users can reduce the surface area of their boards.

Addressing external RAM via the data pointer (DPTR; see the following) works almost exactly the same way, except that instead of _PSEN being asserted during the memory access cycle, either _RD or _WR is asserted depending on whether this is a memory read or write operation. (In the case of a write operation, Port 0, rather than being an input, outputs the desired data during the _WR active period.)

An *extremely* common 8051 reference circuit configuration, particularly for hobbyist experimentation boards, is to have a socket for a 28-pin Flash or EPROM (usually a 27256 32 KB or 27512 64 KB device), a jumper to enable booting off that external memory device, and a single 62256 or equivalent 32 KB static RAM. Another fairly common configuration is similar, but has a single 32 KB Flash configured so that it can be written via the "SRAM side" and read via the "code side" of the external memory interface. By this means, you can have a small bootloader inside the 8051, and a user program inside the Flash memory. The user program can easily be updated using a serial link or some other convenient mechanism.

Of course, you are not limited to connecting just memory devices in this way; you can connect external buffers and latches to expand your I/O capability, or you can add a peripheral expansion IC such as the 82C55. Common practice in such cases is to use external logic; for example, a 74HC138 one-of-eight decoder, to decode a few bits of the demultiplexed address bus and generate the required chip select signals.

The program memory map of the 8051 is relatively simple. There is a single 64 KB program address space from 00000h to 0FFFFh. Depending on the state of the _EA pin, this is either a single external 64 KB space, or a 4 KB internal space and a 60 KB external space (in which case, the lowest 4 KB of the external memory device is not used). Note that code protection features can cause odd results if you're using external code memory. For instance, some 8051 variants, when code protection is enabled, disable the MOVC instruction while the program counter is pointing into external RAM. The rationale is to prevent an attacker from writing a program that reads the low code memory area and outputs it to a port—assuming your design already uses some external code memory, the attacker could simply pull out the device containing the unprotected portion of your program, insert their attack code, and grab the read-out version of your secret code.

Data memory is more complicated. The internal data memory space stretches from 000h to 0FFh. At the bottom of this area, from 000h-01Fh, reside four banks of eight scratch registers. These registers are named R0 through R7; according to the currently selected bank, R0 refers either to location 000h, 008h, 010h or 018h, and so on.

The area from 020h through 02Fh is user-available scratch memory, but has the interesting and useful property that it is addressable at the bit level in a single instruction. The bits in this area are logically numbered from 000h (bit 0 of location 020h) through 07Fh (bit 7 of location 02Fh). When using an assembler that employs standard Intel syntax, you can refer to each location either by its 7-bit address or as "20.0", "2E.4" and so on. It's extremely code size- and speed-efficient to put your program's flags and other bitwise-accessible data in this area, and indeed this is one of the 8051's key strengths.

The remaining area from 030h through 07Fh (the top of internal memory in an 8051) is scratch RAM available for the stack and other user purposes.

Above the internal data memory, from 080h to 0FFh, lives the special function register (SFR) area. These registers control the various hardware features in the microcontroller; the following table shows their functions. Observe that many of these registers are directly bit-addressable. Also note that the SFR area consists mainly of empty holes; these spaces are used to control other hardware in enhanced 8051 variants. For example, since the DPTR is a scarce resource, many 8051 vendors implement an additional DPTR—or even more than one—to speed up operations such as bulk memory copy or to allow the user to maintain state in more than one thread without needing to save and reload the DPTR.

Address	Bit?	Name	Function
080h	Y	P0	Port 0
081h		SP	Stack pointer
082h		DPL	DPTR (data pointer) low byte
083h		DPH	DPTR (data pointer) high byte
087h		PCON	Power Control
088h	Y	TCON	Timer/Counter Control
089h		TMOD	Timer/Counter Mode Control
08Ah		TL0	Timer/Counter 0 low byte
08Bh		TL1	Timer/Counter 1 low byte
08Ch		TH0	Timer/Counter 0 high byte
08Dh		TH1	Timer/Counter 1 high byte

Address	Bit?	Name	Function
090h	Y	P1	Port 1
098h	Y	SCON	Serial control
099h		SBUF	Serial data buffer
0A0h	Y	P2	Port 2
0A8h	Y	IE	Interrupt enable control
0B0h	Y	P3	Port 3
0B8h	Y	IP	Interrupt priority control
0D0h	Y	PSW	Program Status Word (among other things, this contains flags indicating which register bank is currently being addressed)
0E0h	Y	ACC	Accumulator
0F0h	Y	B	B Register

Note that the SFRs can only be accessed by direct addressing modes. For example, "MOV 080H, #012H" will write 012H to the SFR at address 080H, which is to say Port 0. However, "MOV R0, #080H" followed by "MOV @R0, #012H" will not change the state of Port 0.

The reason for this seemingly odd behavior is that 8051 variants with larger internal RAM space—the 8052, for instance—implement RAM "underneath" the SFR area. If you want to access the SFRs, you use direct addressing; if you want to access the RAM with the same nominal address, you use indirect addressing modes.

The final type of memory accessible by the 8051 is "XRAM", which is accessed indirectly using the 16-bit DPTR register and the MOVX instruction (e.g., "MOVX @DPTR,A" stores the accumulator at the external address referenced by DPTR). Just pause here to take stock of all the different address spaces; we have program memory (MOVC, read-only), data/SFR memory, and external data memory (MOVX, read/write).

If you're reading this and beginning to feel that the 8051's architecture is a bit on the rococo side, you're not alone. The 8051's separate (Harvard-architecture) address spaces and, to a lesser extent, the instruction set, are not very friendly to the apprentice embedded developer. (The special 8051s built into ASICs can be even weirder to work with, particularly when you start looking at systems with more than 64K of code memory or XRAM.) The architecture is also significantly unattractive to compilers for high-level languages.

Assuming you need or want to work with the 8051, how do you get started with this core? The usual low-cost method of learning how to use these chips is to buy a single-board computer based on one of the standard parts. In times of yore, it was also necessary to have an EPROM burner, either to burn EPROM-variant microcontrollers directly, or to burn external EPROMs. These days, we have a lot more flexibility in terms of device selection, and the overall cost of entry is much lower.

The lowest-cost method to start working on the 8051 is to build your own development system using a Flash-based microcontroller that doesn't require special programming hardware. In practical terms, this means selecting a part that supports in-system serial programming; these interfaces are usually very simple and easily connected to a PC with simple hardware (in some cases, just a passive cable). I personally enjoy working with the Atmel® AT89S series parts; these are enhanced 8051s running at up to 33 MHz. Atmel provides free in-system programming software (either AT89ISP, or the newer FLIP software) and the schematics for a simple programming cable; you most likely have all the required parts for this in your junkbox.[7] The total cost of building the cable and a homebrew development board is in the neighborhood of $20.

As far as commercial 8051 hardware development tools go, there is a positive cornucopia of choices available to you. The luxury limousine of development hardware is an in-circuit emulator (ICE) such as the Nohau EMUL51-PC. This costly piece of equipment allows you to emulate the target device in real time, trigger breakpoints or other behaviors on a variety of conditions, and more. If you have trace memory installed, you can stop the program execution when a problem is encountered and trace back the program counter to see what your program was

---

[7]   The "official" Atmel programming cable schematic is currently viewable at *<http://www.atmel.com/ dyn/resources/prod_documents/isp_C_v5.PDF>*—however, these deep links tend to move around. Note that this is the same ISP cable Atmel makes/recommends for their ATF15xx family reprogrammable Flash-based CPLDs, although Atmel doesn't presently make that fact clear to the reader.

doing before it went weird. In general, an ICE of this sort gives you unparalleled visibility into your program's state, and it can save you a lot of time.

If a full-speed hardware ICE is beyond your means, some 8051s now have on-chip JTAG interfaces. This sort of interface permits you to perform at least rudimentary debugging using a simple, inexpensive JTAG adapter pod, and appropriate software on the PC side. JTAG solutions are much slower than a full hardware ICE, and by no means as flexible. (For instance, with most JTAG on-chip debugger implementations it's not usually possible to break on accesses to a specific memory address. This is probably the single most useful feature I employ when I have access to an ICE, because it lets me know exactly when a RAM location got corrupted.) However, the JTAG method is inexpensive and undeniably adequate for many applications.

At the bottom of the ease-in-debugging pyramid are the familiar burn-and-pray solutions—either using an EPROM emulator (this is basically a battery-backed-up static RAM, loadable from your PC, that plugs in where you would otherwise insert your external code EPROM), or a reprogrammable Flash-based 8051 device; for instance, the Atmel AT89S family, as I mentioned earlier. If you're heading down that latter path, study carefully exactly what is required to program each chip before you decide what part to use. The Atmel part I just mentioned, for instance, is programmable in-system using a simple hardware interface you can build in a few minutes, and either Atmel's Flash utilities, or the free PonyProg software (available from *<http://www.lancos.com/prog.html>*).

Some other 8051 variants require special, relatively expensive programming adapters. If you don't see "serial ISP" programmability on the device's feature list, chances are you might need to put it in an expensive external parallel programmer in order to reflash it. Some vendors provide low-cost evaluation boards for 8051 type micros; for instance, Keil's list of boards at *<http://www.keil.com/boards>* is quite impressive. The MCB900, for instance, allows you to work on a variety of Philips 8051-type Flash micros, and it retails for around $70. The MCBX51 board works with several vendors' vanilla 44-pin 8051 parts, but it's rather more expensive at just under $300.

In a similar vein, there exists a wide variety of different software development environments available for the 8051. Commercial software packages are available from Keil, IAR, Avocet, Hi-Tech, Raisonance, and others; in most cases, there are free demo versions with various restrictions; Raisonance's evaluation is particularly useful, having a general limit of 4K of object code but otherwise being more or less fully functional.

I personally prefer to use the free compiler package, sdcc (available from *<http://sdcc.sourceforge.net/>*). This is unquestionably the poor man's choice of 8051 C compilers, and it's probably a borderline decision as to whether you'd want to use it in a commercial product due to its relatively inefficient code output (both in terms of size and speed). For large-volume production, a prudent calculation of development tool cost versus the additional per-unit cost of a larger microcontroller will probably lead you away from sdcc. However, most of the 8051 projects I have worked on were written in pure assembly language anyway, so I'm not greatly affected by sdcc's relatively primitive state compared to the commercial C compilers; and in any case, sdcc is perfectly adequate for hobbyist, low-volume and/or educational projects.

Unfortunately, because of the wide variety of special enhancements that various manfacturers choose to add to their 8051 variants, high-level language compilers for the 8051 need to have a lot of detailed knowledge about the target microcontroller. This requires a substantial amount of ongoing support work on the part of the compiler vendor. For this reason (among others), free compilers such as sdcc are unlikely to achieve anything close to the efficiency of a commercial compiler, if only because they won't necessarily "know" about all the different enhanced chip types. For example, simply knowing about the existence of a second DPTR in the target chip could allow a compiler to do a lot of significant time- and space-optimization. Since 8051s tend to find their way into exceedingly cost-constrained applications, unfortunately the limitations of free compiler solutions tend to argue in favor of commercial development tools, which are not cheap. (This commentary doesn't apply to assembly language, of course; if you're hand-writing all your code, then use whatever assembler is convenient.)

In closing, I'll share an interesting point, which you might find to be food for thought: One of the most respected manufacturers of 8051 development tools is Keil. In 2005, Keil was acquired by ARM. There has been rampant speculation about the possible ramifications of this. Some people point to the fact that Keil has experience developing very sophisticated compilers for 8051, and theorize that ARM will exploit this to develop ARM compilers of greatly improved efficiency. Others point out that 8051 is a very mature product, and the cost-performance curves of high-end 8051 and low-end ARM almost overlap. There are an awfully large number of 8051 applications that are starting to creak at the seams as the list of mandatory "fashionable" features (wired Ethernet, WiFi®, Bluetooth® and Web-enabled control functions, to name a few of the most popular) becomes more convoluted. These big applications are ripe for migration to single-chip ARM parts. Hence, there is some speculation that ARM might conceivably reconfigure Keil's product line in order to encourage people to migrate up to ARM-cored parts. This could be the thin end of the wedge; ARM might sweep downward through product portfolios, starting at the biggest and most complex, but eventually supplanting 8051 in even the simpler applications.

3.3   Atmel AVR®

Although technically in the same genus as the 8051, the Atmel AVR is a considerably friendlier architecture. Before going any further in this section, I had better come clean about my addiction; I really *love* the AVR family. It is definitely my favorite 8-bit platform. I've found that it is both easy to work with and inexpensive to get tooled up for this micro. It's very well-suited for hobby projects and the low-volume commercial applications with which I occupy my spare time. (You will notice that large-scale commercial applications are missing from that list. I've been trying to find an excuse to use AVRs in my day job; although management is cautiously enthusiastic, the sole factor that has precluded a migration to AVR so far is the existence of large amounts of legacy assembly language for other cores.)

Now that all these virtues have been extolled, full disclosure requires me to point out some downsides. The first, obvious downside is that AVR is a proprietary core, and hence all AVR parts are single-sourced from Atmel. By itself, this is not such a terrifying prospect; virtually every microcontroller is a single-source part. However, attached to this comment is the fact that Atmel has a less than perfect track record for dealing with small customers. If you browse through the archives of comp.arch.embedded, you'll find periodic complaints from people about Atmel's vaporware products, strange and arbitrary product discontinuation decisions, the utter impossibility of obtaining samples of any new part (I've experienced this problem firsthand, and can vouch for it), difficulty with sudden jumps in product lead time, and inaccurate estimates on delivery schedules. As a result of this, it is prudent to eschew brand-new parts for a while after they appear on the Atmel product linecard. It is also a wise idea to design out parts as soon as they migrate to the "not recommended for new designs" status unless you're a big enough buyer to merit a last time buy letter. If you're relying on low-volume distribution and you don't keep track of the product lifecycle, you might have an unpleasant surprise next time you go to order parts. As a general rule, if you stick with high-volume variants that are readily available at retail from stockists like Digi-Key, you are fairly unlikely to have a problem.

As a specific example of the AVR's capabilities, let's consider the ATmega32L in DIP-40 packaging illustrated in Figure 3.2.[8] The ATmega32 is one of the higher-end AVRs, though not the largest part by any means. (Note that there are a few slightly different flavors of AVR; very small parts, mid-range parts, and large parts. The mega32 is one of the large parts. The larger cores have an enhanced instruction set that is a superset of that found in the smaller cores.)

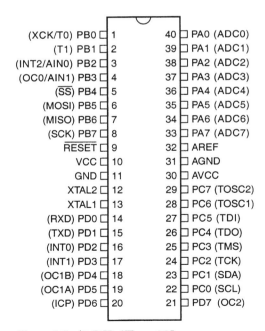

(XCK/T0) PB0	1	40	PA0 (ADC0)
(T1) PB1	2	39	PA1 (ADC1)
(INT2/AIN0) PB2	3	38	PA2 (ADC2)
(OC0/AIN1) PB3	4	37	PA3 (ADC3)
($\overline{SS}$) PB4	5	36	PA4 (ADC4)
(MOSI) PB5	6	35	PA5 (ADC5)
(MISO) PB6	7	34	PA6 (ADC6)
(SCK) PB7	8	33	PA7 (ADC7)
$\overline{RESET}$	9	32	AREF
VCC	10	31	AGND
GND	11	30	AVCC
XTAL2	12	29	PC7 (TOSC2)
XTAL1	13	28	PC6 (TOSC1)
(RXD) PD0	14	27	PC5 (TDI)
(TXD) PD1	15	26	PC4 (TDO)
(INT0) PD2	16	25	PC3 (TMS)
(INT1) PD3	17	24	PC2 (TCK)
(OC1B) PD4	18	23	PC1 (SDA)
(OC1A) PD5	19	22	PC0 (SCL)
(ICP) PD6	20	21	PD7 (OC2)

Figure 3.2 40-DIP ATmega32L.

---

[8]   Information in this section is taken from the ATmega32/ATmega32L datasheet, document #2503F-AVR-12/03.

As with most other micros, the pinout is again dominated by GPIOs; ports A through D are 8-bit general-purpose ports, each pin of which can be individually programmed as inputs or outputs. The following are the other pin functions (including those multiplexed onto GPIO ports).

- *XTAL1, XTAL2* – Crystal or ceramic resonator connections. The AVR can use an internal RC oscillator clock source, an external RC, or an external resonator.

- *AVCC, AREF* – Vcc and reference voltage for the on-chip ADC.

- *ADC0* through *ADC7* – Analog-to-digital channels. Each line can be individually assigned to the ADC module.

- *XCK* – External clock input for the USART module. This is used if you are implementing a synchronous serial protocol with the clock sourced from some external device.

- *T0* – Timer/Counter 0 external input.

- *T1* – Timer/Counter 1 external input.

- *AIN0* – Positive input for on-chip analog comparator.

- *INT0, INT1, INT2* – External interrupt 0, 1 and 2 inputs.

- *AIN1* – Negative input for on-chip analog comparator.

- *OC0, OC2* – Timer/Counter 0 and output compare match (used for PWM drive applications).

- *SS* – SPI Slave Select input (when operating as a SPI slave device).

- *MOSI, MISO* – SPI bus master out/slave in and master in/slave out pins.

- *SCK* – SPI serial data clock.

- *SCL, SDA* – Two-wire bus (I^2C by another name) clock and data lines.

- *TCK, TMS, TDO, TDI* – JTAG debugging interface pins.

- *TOSC1, TOSC2* – Timer Oscillator crystal resonator lines. Timer/Counter 2 can be operated in an asynchronous clocking mode where the clock source is provided by an external resonator connected to these pins.

- *RXD, TXD* – USART receive and transmit data lines, respectively.

- *OC1A, OC1B* – Timer/Counter 1 Output Compare A and B match outputs.

- *ICP* – Timer/Counter 1 Input Capture Pin.

Note the rich peripheral set on this part. If you're contrasting it against the 8051 I discussed in the previous section, be advised that it is not a fair apples-to-apples comparison. The 8051 I described earlier is *strictly* a basic, generic part; I chose it specifically because this is a lowest-common-denominator chip available from several vendors. There are certainly 8051 variants with every peripheral you'll find in the AVR family; you just need to shop around. Much like the AVR, these more enticing chips will be single-source parts, or nearly so.

Power consumption on the AVR family is in the middle to low end of 8-bit parts. Most devices offer selectable power consumption options. The exact power-management features available depend on the device, of course, but most of them center around keeping the device in sleep until an interrupt occurs. As an interesting side note, in order to obtain the clearest possible results in ADC samples taken by the device, it's absolutely necessary to use these power-down modes so the CPU core isn't injecting noise into the analog reading.

At the time of writing, Atmel has just announced the impending release of a new "picoPower" series of AVR devices. These are standard AVRs with a reduced power requirement, identified by a P suffix on the part number; the first to be offered is the Atmega169P. Operating at 1.8V, this device draws 330 µA at 1 MHz, or as little as 10 µA when operating from a 32.768 kHz watch crystal.[9] Even with an active LCD display, this number only rises to 25µA. These devices are aimed very clearly at the MSP430's market space, and in fact Atmel's marketing literature makes direct comparisons.

AVR is a Harvard architecture, with two main address spaces—program and data memory. Program memory stretches from $0000 to $3FFF (32K of space); data memory stretches from $0000 to $085F. The data space from

---

[9] These numbers are 350µA and 20µA respectively for the older non-picoPower ATmega169; with LCD active, the part draws 40µA.

$0000–$001F is reserved for general-purpose registers; $0020–$005F is occupied by I/O registers. The 2K of on-chip RAM resides from $0060–$085F. From a superficial look at the instruction set, you might be led to believe that there is a separate I/O space, since there are dedicated IN and OUT instructions that operate on the I/O register area. This isn't the case; IN and OUT are just convenience features.

The ATmega32 also has a third address space, EEPROM memory—this is, however, not directly addressable by the core. In a similar fashion to most other parts with on-chip EEPROM, this memory is accessed through an 8-bit window; you write the desired EEPROM address to a pointer register and access data through another register.

Speaking of address spaces, observe that this particular AVR variant does not have external address or data buses (although some do). Since AVR cannot execute code out of on-chip RAM, this means that you can't attach external removable code modules (software cartridges) and have them directly mapped into the code space as an executable program. You also can't add more directly addressable RAM. If either of these functions is going to be important to your application, you should either look at one of the larger AVRs that does offer external buses, or (in the expansion cartridge example) consider writing an interpreter that can load and execute programs off external serially accessed Flash chips.

The AVR core has 32 8-bit general purpose registers, named R0 through R31. These registers live in the bottom 32 bytes of the data address space. The instruction set is not entirely orthogonal with respect to these registers; registers R16 through R31 are treated slightly differently than R0 through R15. Apart from this detail, however, the instruction set operates more or less equally on all the general-purpose registers.

The top six registers have an additional special function; they can be treated as three 16-bit registers for the purpose of certain instructions involving indirect addressing modes and 16-bit arithmetic. When you are taking advantage of this capability, these 16-bit registers are named X (R26, R27), Y (R28, R29) and Z (R30, R31). In each case, the lower-numbered register contains the lower-order bits of the 16-bit concatenation. (Observe that this is just a nomenclature detail. You can access the individual 8-bit registers at any time and they will contain valid data.) From the point of view of high-level language usage, observe that the limited number of 16-bit registers acts something like the DPTR restriction in the 8051 family; it's something of a bottleneck.

There are a couple of slightly odd things about the AVR family, and both of them relate to programming the parts. The first strange thing is that you have to program the device's three memory areas—program memory, EEPROM and fuse bits—separately. There is no universally recognized way to combine all this data into a single file for one-step programming. The way I usually work around this is by using a command-line programming tool (avrdude, to be exact; the homepage for this open-source tool is *<http://savannah.nongnu.org/projects/avrdude/>*); I can then embed a one-step programming option into my project's makefile.

The second oddity is that the internal RC oscillator requires some device-specific calibration in order to operate exactly on its nominal frequency. This is not, of itself, terribly odd for on-chip RC oscillators in microcontrollers. What is unusual is that the factory calibration bytes are stored in such a way that they can't be used directly; they are located in the device signature area, which can only be read by external programming hardware.[10] You can select RC clock speeds of 1, 2, 4 or 8 MHz in the fuse bits; your software then needs to load the correct device-specific calibration byte into the OSCCAL register at powerup.

This process is handled automatically for the 1 MHz case; at reset, OSCCAL is automatically initialized with the 1 MHz calibration value. If you're using one of the faster speeds, Atmel has thoughtfully provided the correct calibration constants for you in the chip, but there's no way your software can get at them directly.[11] The recommended workaround for this is to read out the signature area with your programming hardware, retrieve the appropriate calibration byte for the speed of interest, and write it into a spare byte of software-accessible EEPROM. This has to be done individually for each chip you program; the calibration constants are dependent on process variations. Your software then needs to read the relevant EEPROM byte at power-on reset, and write that value into OSCCAL. This is really quite annoying; I wish Atmel had either made the part auto-load the correct calibration byte for each RC oscillator mode, or at least provided a method for software to access the factory calibration area.

---

[10] The device signature area appears to be merely a reserved area of EEPROM cells. It contains factory calibration values and the device signature bytes used by programming hardware to determine the chip version.

[11] As this book was going to press, Atmel was working to eliminate this particular oddity.

There are quite a few different development hardware options for working with the AVR. Atmel's very lowest-end development board (really, more of a demonstration than a development board) is the AVR Butterfly, which sells for about $20 but is frequently given away as a freebie at Atmel seminars and "lunch and learn" events. The Butterfly consists of an ATmega169 microcontroller with a small alphanumeric segmented LCD, lithium coin cell, a 4-way miniature joystick of the type found on cellphones, 4 Mbits of off-chip Flash memory, a temperature sensor (thermistor), a light sensor (CdS cell) and a piezo speaker element. The board has a lapel clip so you can attach it to your clothing and use it as a name-tag badge. It is shipped preloaded with some demo software, including a bootloader—so it can be reprogrammed over a regular serial port; you don't need to own Atmel-specific programming hardware. However, the board does have a JTAG connector and you can use it as a low-cost evaluation platform for the ATmega169.

The next step up from the Butterfly is the STK500 development board, again direct from Atmel, which retails for about $79. This is the recommended entry-level evaluation board for the AVR series, and it's the board I'd recommend you acquire if you're just starting out with these parts. The STK500 connects to your PC over a standard serial interface.[12] It has sockets for most of the DIP-package AVR variants, and you can run some simple applications directly on the board. Note that plug-in adapters are also available so that you can use the STK500 to work with some of the larger AVR parts that don't come in DIP packaging.

The STK500 has eight LEDs and eight pushbuttons, as well as headers to allow you to connect to all of the I/O lines of the micro. You can simply put your chip in the appropriate socket and connect the rest of your circuit to the headers. Using a (supplied) 6-pin cable, you can also use the STK500 to perform in-system programming of AVR devices on your own custom boards. The STK500 supports serial (SPI) programming as well as parallel programming; the advantage of this is that if you accidentally switch off the SPI programming mode in the fuse bits of your target device, you can use the STK500 to recover and reprogram the chip, as long as you can remove the device from your circuit.

---

[12]   The STK500 board plays well with most of the USB-to-serial adapters I have tried under Windows, Linux and Mac OS X. This is an important point in these dark days of legacy-free PCs. Most of the homebrew serial-connected device programmers don't work properly over USB converters.

The one thing the STK500 doesn't do for you is offer debugging support. It's strictly useful for burn-and-pray type development. If you want to debug your code, you need either a full ICE (for the low-end parts) or a JTAG ICE for the higher-end parts that feature JTAG interfaces. Atmel's JTAG-ICE is a serial-connected device, and fairly expensive; it's significantly cheaper to buy Olimex's JTAG-ICE clones, which are available in both serial and USB flavors. (The USB flavor is simply the serial version with an FTDI USB-to-serial converter chip built into the housing.) Olimex can be found at *<http://www.olimex.com/>*—they make all sorts of inexpensive evaluation hardware, as well as offering low-cost PCB prototyping services. Their off-the-shelf products are distributed in the U.S. by Spark Fun Electronics.

Atmel's newer parts also feature an exceedingly nifty one-wire debugging interface called *debugWire*; this interface allows you to debug your code using only the ground and reset lines, thereby allowing you to use the JTAG pins as I/O ports. debugWire is supported by the JTAG-ICE Mk. II, which has replaced the original JTAG-ICE.

From a software perspective, it's easy to get started on the AVR in assembly language using Atmel's free AVR Studio® software (available for Windows® only). There are actually three different assemblers available under Windows; Atmel's own, Tom's Linux AVR Assembler "tavrasm" (available from *<http://www.tavrasm.org/>*; despite the name, you can build and run this assembler happily on Windows using Cygwin) and the AVR-targeted flavor of gas, the GNU assembler. If you're going to do your programming in assembly language, I'd advise either tavrasm or the Atmel assembler; although gas is perfectly functional, it won't directly assemble off-the-shelf code that was written for Atmel's environment. tavrasm, on the other hand, is specifically designed to work exactly like the Atmel assembler.

Note also that AVR Studio is a fully integrated IDE comprised of a source editor, assemblers, and software to talk to AVR programming and debugging hardware. (A software simulator is also included.) If you use a different assembler—for instance, if you're working inside Linux or Mac OS®—you'll need separate software to burn chips. The best choice right now is avrdude, which I mentioned earlier. From time to time, Atmel issues updated firmware for their hardware such as the STK500, the JTAG-ICE and the AVRISP programming cable. These updates are shipped inside newer versions of AVR Studio; when you connect to the development hardware, the IDE checks the firmware version and automatically updates older boards. Sometimes, these updates involve radical protocol changes that completely break all older software. avrdude is the most up-to-date open-source programming utility; it's well-supported and won't remain broken for long if Atmel change something. Other alternatives, such as the once-popular uisp program (*<http://savannah.nongnu.org/projects/uisp/>*), have fallen behind in their updates and no longer work with current development hardware.

The AVR's fairly neat architecture and relatively tidy instruction set make it reasonably friendly to C compilers; certainly much more friendly than the 8051. In fact, the basic architecture was designed (by two Norwegian university students) with the specific goal of efficient C code execution. Runtime performance is good, but compiled object size is generally not fantastic, particularly when dealing with 16- and 32-bit arithmetic. There are several extant C compiler packages targeting the AVR; the two with which I am most familiar are gcc and a commercial product from Rowley Associates *<http://www.rowley.co.uk/>*. Rowley's product is a full IDE-style compiler/debugger which is simple to use and provides turnkey operation. It is available for both Windows and Linux.

A gcc installation is, of course, composed of several components; GNU binutils, gcc itself, a C runtime library for the AVR (avr-libc, available from *<https://savannah.nongnu.org/projects/avr-libc/>*), perhaps the GNU debugger gdb, and a programming tool such as avrdude, not to mention a sourcecode editor. I prefer to use Eclipse *<http://www.eclipse.org/>* as my sourcecode editor, as it provides me with a consistent user interface for building all my projects—Linux, Mac OS, AVR, ARM, and so on. Note that there is a prepackaged suite of ready-to-run GNU tools, referred to as WinAVR, available from *<http://winavr.sourceforge.net/>*. I would normally download the individual components and build them myself rather than relying on a prepackaged installer, but this is to a large extent a matter of individual taste.

Out of the available build environments, I generally prefer gcc, although it does not yield the smallest or fastest code. Since I work with gcc on numerous platforms, I have less porting work to do if I standardize on gcc wherever it is available. This is important to me, because time is valuable; I need to be able to write code on (say) AVR and use it on (say) ARM with the minimum possible effort. By using the GNU toolchain throughout, in a best case all I need to do is change the makefile.

A final note on development environments: If you're a Mac OS user by preference (OSX only, on Intel or PowerPC), you should be aware that the AVR is by far the easiest microcontroller family for you to use as a learning tool. All the open-source software tools compile and run on OS X without any sort of drama. The inexpensive hardware development tools are connected either over serial ports or internal USB-to-serial converter chipsets; at most, all you'll need to do is find a $10 USB-to-serial cable in order to get up and running. For most other microcontrollers, you'll have difficulties of one kind or another working on Mac OS. Getting a working toolchain built is not usually a problem, but finding a way to get your compiled code into the evaluation board or your own circuit can be annoyingly difficult since a Mac has no parallel ports, and Linux drivers for USB-hosted emulators can't be used in Mac OS.

3.4   Texas Instruments MSP430

It's slightly cheating to lump the MSP430 in with 8-bit microcontrollers, since it's a pure 16-bit device, but it's sold into the same sorts of applications as 8-bit micros, and hence belongs in this chapter despite the fact that the part's classification is not completely semantically aligned with the other products in the lineup.

Compared with, say, the 8051, the MSP430 is a relatively young family—it's only about ten years old. The early variants were developed for low-cost, low-power measurement applications. The design goals for these applications have been carried into the current range of products: the MSP430 is a low-power-consumption family with the usual selection of interrupt-woken low-power snooze and sleep modes, and it has a very flexible programmable analog-to-digital converter module. Some parts also feature on-chip LCD controllers. The parts feature a programmable DCO-based clock generation module that can clock all sections of the chip from a standard 32.768 kHz crystal.

Now, it's an obscure but well-recognized ritual among engineers and computer scientists to gauge the architecture of any new device on the basis of its similarity to the PDP-11.[13] The strongest term of approbation you can use for a CPU design is to say "It's just like a PDP-11!" Exactly why this is universally regarded as a Good Thing is not exactly clear, but in any case, this epithet is frequently applied to the MSP430. (I've always felt that this is the same sort of statement as saying "My 2007 Mercedes convertible is just like a 1965 International Harvester Scout light truck. They both have pneumatic tires and a removable roof!") What I think these people probably mean is that the MSP430 has a very nice orthogonal instruction set and simple memory addressing scheme.

Flash programming and software debugging for most MSP430 parts is carried out using an on-chip JTAG interface. TI sells a relatively low-cost kit containing their "FET" (Flash Emulation Tool, a simple parallel-port-based JTAG pod), and a free version of the IAR toolchain. Other vendors have cloned this JTAG pod; Olimex, for instance, sells a completely compatible third-party version for even less than the cost of the FET. Texas Instruments and Olimex, as well as Rowley, all offer USB-connected JTAG debuggers, which operate much faster than the cheap and nasty parallel port units.

As a breaking-news item, just as this book was going to press, TI released an ultra-low-cost USB-connected demo/development board called the *eZ430*. This is a tiny device, looking very much like a USB Flash drive. An even smaller target board is connected to a header on the end opposite the USB connector. Currently, TI only sells the eZ430 as part of the eZ430-F2013 kit, which contains the eZ430 emulator and a target board with an MSP430F2013 processor, space for a 14-pin, 100 mil header to bring out all the F2013's pins to your prototype, and a single LED. (By the way, the USB Flash Emulation Tool built into the eZ430 supports all MSP430F20xx series parts; the MSP430F2013 is simply the highest-end member of that family, with 2K Flash, 256 bytes of info memory and 128 bytes of RAM.) The retail price of this kit is $20, though as with Atmel's Butterfly, the kit is a free giveaway at Texas Instruments seminars. If you want to learn the MSP430 instruction set and experiment with the platform in general,

---

[13]   If I ever design a microprocessor, I am going to print a line drawing of a complete PDP-11/20 front panel on the top of the package.

the eZ430 is probably the best turnkey solution currently available. Note that it is *not* a generic JTAG tool; it uses the two-wire "Spy Bi-Wire" debug protocol and it currently only works with a few select devices. Spy Bi-Wire is clearly an attempt by TI to respond to Atmel's two-wire debugWire protocol; it's very useful not to have to waste numerous I/Os on the JTAG interface, and I look forward to Spy Bi-Wire being offered on all of TI's parts someday.

There are a couple of disadvantages of the MSP430 from the point of view of the hobbyist or one-person contract shop. The first is that the parts have a 3.3V I/O voltage with no 5V tolerance; not entirely unexpected, since this family is specifically aimed at low-power applications. This low I/O voltage is not an utterly terrifying fact, but it does mean that interfacing to typical hobbyist projects (motor controllers, for instance) is potentially a little bit more complicated than it would be for a 5V device. Of course, this low-voltage design stems from the fact that the MSP430 is intended for battery-powered or otherwise power-constrained applications.

A much more irritating issue for someone trying to get up and running quickly is that the entire range of MSP430 family devices (it's not a particularly large range, by the way) is only available in fine-pitch surface-mount and leadless package variants. This makes it difficult to put together a hand-built prototype on matrix board. The only real answer to this is either to build a custom PCB for your application, or buy off-the-shelf evaluation hardware. Fortunately, this is not as expensive as you might think; Olimex sells two flavors of bare-bones evaluation boards for several different MSP430 devices.

The devices that Olimex calls *proto boards* (presently available for MSP430F1121, 123, 1232, 149, 169, 1611 and 2131) consist of the MSP430 device in question, a fairly generous prototyping area with plated-through holes on 0.100" centers, and some useful infrastructure such as a power supply, serial driver and standard 9-pin D connector, 32.768 kHz clock crystal, JTAG header and reset button. The so-called header boards are simply the chip by itself on a tiny board with 0.100" headers around the edge bringing out all the pins. You can plug this header board directly into a prototyping board, wire-wrap to it, or solder it onto a piece of matrix board (although I'd prefer to socket it for reuse, myself). The header board also contains a 32.768 kHz crystal, a JTAG header, and a socket for a high-frequency crystal, if you wish to use one. The price difference

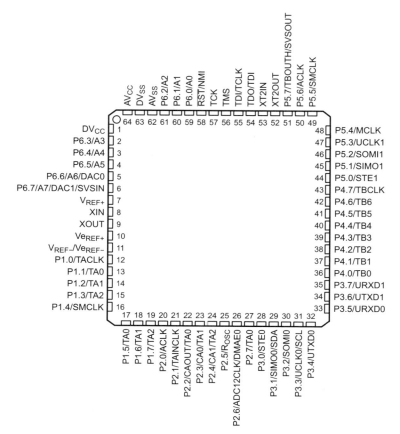

Figure 3.3 MSP430F169 pinout.

between the header boards and the bare chips in single-piece quantities is about 2x to 2.5x, which is really very reasonable.

Let's look a little closer at a specific part, the MSP430F169, shown in Figure 3.3. This is one of the highest-end parts in the family, with 60K of Flash memory, 2K of RAM and 48 I/O pins in a 64-pin package, either LQFP or QFN. We'll begin, as before, by looking at the pinout.

Due to the high pin count, I won't break down all the functions here; I'd just like you to observe the major features of the microcontroller:

- Six 8-bit GPIO ports, with other functions multiplexed in the usual way. Ports 1 and 2 have selectable-polarity interrupt capability.

- Hardware multiplier for 16x16, 8x8, and 16x8/8x16 operations in a single instruction cycle. This allows some simple DSP type MAC operations to be implemented in a speed- and power-efficient manner. Not all MSP430s have this feature; it's a peripheral, not part of the basic processor core.

- Two USARTs. Both USARTs support SPI and regular asynchronous serial communications. USART0 also supports high-speed I²C (real, officially licensed, Philips-compatible I²C at that), over DMA if desired.

- One 16-bit timer_A3 module (more on this later) with three capture/compare registers.

- One 16-bit timer_B7 module with seven capture/compare registers.

- One comparator_A analog comparator module.

- ADC12 (12-bit ADC) module supporting eight external analog channels as well as an on-chip temperature sensor. You can also sample the external voltage reference pins. TI's ADC module is quite extraordinary in terms of the flexibility it offers you; it can be connected to the DMA module to sample without CPU intervention (the CPU can even be sleeping to reduce power consumption) and the actual conversion mechanism can be tweaked to an amazing level in software. This can save you external level-matching components.

So, what are all these names like timer_A3, comparator_A, and so forth? The MSP430 family is very modular internally, and Texas Instruments makes this very clear in their documentation. Each possible peripheral module that can be attached to an MSP430 core has a name representing its function; the datasheet for an individual chip tells you what modules the chip contains, and the cores themselves are documented in a generic manual that's not specific to any one MSP430 device. (This generic manual is Texas Instruments' document #SLAS368D, "MSP430x15x, MSP430x16x, MSP430x161x MIXED SIGNAL MICROCONTROLLER.") Because of this explicit similarity, it's unusually easy to move your code between different members of the MSP430 family. As long as the device to which you're migrating contains the peripheral you want, all you need to do is make sure the I/O mapping is up-to-date and recompile the code.[14]

---

[14] AVR's intra-family similarity is quite good, but MSP430's is really excellent.

The internal architecture of the MSP430 core is beautifully neat and tidy, as I hinted previously (though some people do complain that the instruction set architecture is billed as RISC, but implemented as CISC). There are sixteen general-purpose 16-bit registers, named R0 through R15, which are all treated identically from an instruction set perspective. Although I call them "general-purpose" (since the datasheet does so), the first four have defined functions; R0 is the program counter, R1 is the stack pointer, R2 is the status register and R3 is the constant generator. This whole system is very C-friendly and enjoyable from a programming standpoint.

The MSP430 core has a traditional von Neumann architecture. Some references quote this as a disadvantage, which is rather incomprehensible since there's no obvious characteristic of the micro, compatible with its design goals as a low-power device, that would be improved by switching to a Harvard architecture.

Source	Dest	Name	Syntax	Meaning
Yes	Yes	Register mode	Rn	Contents of Rn is operand.
Yes	Yes	Indexed mode	X(Rn)	Memory at (Rn + X) is operand.
Yes	Yes	Symbolic mode	ADDR	Memory at (ADDR) is operand, with relative addressing. Note that this is actually encoded as X(PC).
Yes	Yes	Absolute mode	&ADDR	Memory at (ADDR) is operand, with absolute addressing. Note that this is actually encoded as X(SR) (i.e., a constant offset from the zero constant generator).
Yes	No	Indirect register mode	@Rn	Memory at (Rn) is operand.
Yes	No	Indirect register mode with autoincrement	@Rn+	Memory at (Rn) is operand. Rn is incremented after instruction (by 1 for byte-wide operations, by 2 for word-wide operations).
Yes	No	Immediate mode	#N	Immediate 16-bit constant is source. This is implemented as @PC+ (i.e., a special case of indirect register mode with autoincrement).

Code memory, data RAM, "info memory" (two small, identical-length in-application-programmable Flash blocks used for storage of parameters in the same way you would use the EEPROM module on other microcontrollers) and I/O all reside in a single 16-bit address space. The orthogonality of the instruction set is quite amazing; is a list of the addressing modes is shown in the preceding table.

Any instruction can use any valid combination of addressing modes. For instance, it's perfectly legal to use a register's contents as the destination of a CALL instruction, as in CALL R5. Even better, you can use CALL X(R5) to route program execution flow through a jump table. All this flexibility does take some getting used to if you're coming from other microprocessors with a more traditional (i.e., less flexible) programming model, especially bizarrely constrained architectures such as PIC®. The memory map is likewise laid out in a very tidy manner.

Range	Contents
0x0000–0x000F	8-bit special function registers.
0x0010–0x00FF	8-bit peripherals.
0x0100–0x01FF	16-bit peripherals.
0x0200–0x09FF	2 Kbytes of RAM.
0x0C00–0x0FFF	1 Kbyte bootloader memory (ROM).
0x1000–0x10FF	256 bytes of information memory, in two 128-byte blocks. This is software-programmable nonvolatile memory used for storing calibration constants and similar data.
0x1100–0xFFDF	60 Kbytes (less 32 bytes) code Flash space.
0xFFE0–0xFFFF	Interrupt vector area (code Flash).

"Information memory" works a little differently from the nonvolatile storage peripherals available in most other 8-bit micros, and you might find it a little irksome for some applications. It's implemented as two 128-byte Flash blocks[15] named segment A and segment B. Erasure can only be performed with block granularity. Reads and writes can be performed with either byte or word granularity, with the caveats that (a) you can only turn "1" bits to "0", not the reverse; and (b) each time you write to any location in a block, it eats into that block's accumulated programming voltage time. If this cumulative programming time gets too large, you'll need to erase and rewrite the entire block. Of course, there is an overall erase-write lifetime for the memory array, also. The downside of this system is that it's relatively annoying to implement nonvolatile data that are updated frequently (vending machine audit counters, powerup lifetime timers, and so forth), because you need to read the entire segment into RAM, update the value you wish to change, erase the Flash version, and write the entire segment back out. The reason that information memory is divided into two blocks is so that you can implement redundant storage schemes to guard against a power failure or other glitch during the erase or write operations.

Contrast this with the traditional on-chip EEPROM found on most other parts in the same application space; the traditional method allows you to read, erase and write bytes individually (though it's not quite as easy to work with those devices, since they don't usually map the entire storage array into the processor's address space). If your application needs to store both frequently updated data (such as audit counters) and infrequently updated data (serial numbers, passwords and so forth), design your storage algorithm with great care, taking the MSP430's limitations into account. Also look very closely at the Flash write endurance for your selected part and compare it with the frequency with which you're erasing and rewriting the information memory.

The on-chip bootstrap loader (BSL) is a handy ROM-based feature allowing you to reprogram the Flash memory using a three-wire serial interface; P1.1, P2.2 and ground. Using the BSL is documented in Texas Instruments document

---

[15]   Some MSP430s split info memory into 64-byte blocks. Most devices have the dual 128-byte banks, however.

#SLAA089, "Features of the MSP430 Bootstrap Loader"; in brief, you can reprogram the chip a block at a time using a simple 9600bps serial protocol, which can be controlled from a host PC or another external embedded device. The MSP430 bootloader can also be password-protected to prevent random people from peeking into your code.

If you're going to program the MSP430 in assembly language, then the limited IAR Workbench provided for free by Texas Instruments is more than adequate for smaller projects. As long as you're working in assembly language and using Windows as your development platform, I can't see any obvious reason why you would want to upgrade to anything more advanced.

However, several C compilers are available for the MSP430, and there are significant differences between them. There is obviously a full-version product from IAR; if you have been using the limited version, then upgrading to this full version will leverage your toolchain-specific experience as much as possible. I have to say, however, that I've had nothing but disappointing experiences with IAR; almost everything negative I have to say in Section 3.8 is applicable to my history with IAR's products, so I couldn't recommend that you go this route.

Alternatively, similar to their product for the AVR family, Rowley has a version of their CrossWorks assembler/C compiler/IDE that targets MSP430. It's quite intuitive to use once you spend a day or two familiarizing yourself with all the options, and the price and licensing conditions are much less onerous than IAR's product; the code that's generated is comparable between these two packages. Furthermore, there's not a lot of work you need to do in order to port IAR-specific sourcecode to CrossWorks; the syntax for accessing compiler-specific features is very similar. CrossWorks supports the Texas Instruments FET (and the third-party clones from various vendors) as well as several other pieces of debugging hardware. Rowley also offers their own USB debugger.

For my personal projects, however, once again I prefer to use gcc. The MSP430 flavor of gcc (and associated tools) is available from *<http://mspgcc.sourceforge.net/>*. You should make sure to read the FAQs for this product at *<http://mspgcc.sourceforge.net/ faq/>*—there are some subtleties that you might not immediately appreciate, especially if you're not accustomed to using a high-level language in an embedded environment.

To summarize, then: MSP430 is an easy-to-understand, very flexible and fairly high-performance architecture designed specifically for power-critical applications. The main downsides are that the device family is not particularly large (meaning you might have to buy more micro than you want, simply in order to get a specific feature) and the parts are not perfectly suited to hobbyists because of the relative difficulty of prototyping. However, the parts are popular in both commercial and hobby applications, particularly where battery- or solar-powered operation is essential.

3.5   Microchip® PICmicro®

Microchip's PICmicro[16] microcontroller offerings are many and varied. At the low end, they offer tiny six-pin SOT23-package PIC10F devices for applications like glue logic replacement or generating reset signals; at the high end, they have the forthcoming 16-bit PIC24 series parts. (Somewhere in the mix are the dsPIC DSP devices, but these aren't really pure PICs so we won't consider them here.)

Many embedded codehounds, myself included, have worked extensively with PICs on both hobby and commercial projects. In fact, I worked with PICs before I ever touched MSP430 or AVR devices; you can still see archived pages for a couple of my ancient hobby PIC projects (Picxie and Picxie 2) at *<http://www.zws.com/products/>*. Besides making their way into high-volume appliances (Visteon,[17] for instance, was at one time reputedly a big consumer of PICs; several variants are widely used in remote-entry wireless keyfob transmitters), Microchip's PIC offerings are also used fairly extensively for teaching purposes.

Let's look at an exemplar PIC part and contrast it against the others we've examined so far. Figure 3.4 is the pinout for the PIC16F84A[18] in 18-pin DIP or SOIC package.

---

[16]   PIC is an acronym for Programmable Interface Controller or Programmable Intelligent Computer, depending on how far back in history you go. Microchip doesn't actually talk about the name being an acronym, however. Although everyone calls them PICs, they are technically PICmicros per Microchip's official terminology.

[17]   A spinoff from the Ford Motor Company. Visteon develops automotive electronics, mostly focusing on entertainment and navigation equipment.

[18]   Please note that this specific part is not recommended for new designs. The PIC16F627 or PIC16F628 are more modern choices; however, most of what I say here still applies. Besides, the PIC16F84A was so amazingly popular that it simply refuses to die!

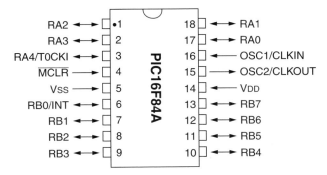

Figure 3.4 PIC16F84A 18-DIP pinout.

The pin descriptions are simple enough:

- *RA0* through *RA4* are the five accessible bits of an 8-bit GPIO port.

- *T0CKI*, multiplexed on RA4, is an external clock input for Timer 0.

- *RB0* through *RB7* form a second 8-bit GPIO port with a programmable weak pullup feature. The top four bits of this port have interrupt-on-change capability. RB6 and RB7 are also used for in-system programming; RB6 is the clock line, and RB7 is the data line.

- The *INT* input, multiplexed on RB0, can be used as a general-purpose external hardware interrupt line.

- *_MCLR* is the external active-low reset input.

- *OSC1* and *OSC2* are used to attach an external crystal or ceramic resonator, an RC oscillator, or an external digital clock source.

Obviously, this is a much smaller part, both in terms of program size and I/O budget, than the others I've discussed so far. It also doesn't offer terribly much in the way of on-chip peripherals. I chose this chip partly because I'm familiar with it, but mostly because it has historically been one of the most popular PIC devices used by hobbyists. You'll find quite literally *thousands* of projects based around this microcontroller! Also note that the comparisons I'll make here should not be dismissed completely as an artificial apples-to-oranges fabrication, because the other three core families I've discussed in this chapter all scale down to approximately the size of the PIC16F84A (or even smaller) without exhibiting the same oddities. For example, 8051s go down to 1K of code in an

8-pin SOIC package (Philips P89LPC903, for example) and still keep almost exactly the same programming model. The MSP430F2001 is a 1K Flash part in a 14-pin package, and it still has the full, elegant MSP430 core and memory architecture. Thus, small size alone does not excuse weirdness.

So, what oddities does PIC exhibit, you ask? There's quite a list. To begin with, most of the "8-bit" PIC variants, including this one, actually have a 14-bit instruction word. (There are several 8-bit cores available; a 12-bit-word core, a 14-bit-word core, and two 16-bit-word cores, the PIC17 and PIC18 series. The idiosyncrasies are different among these families.) Microchip always quotes their code size in words, so a 1K 14-bit PIC is actually 1.75 kilobytes of actual Flash space. This isn't really a big deal unless you're trying to parse the raw HEX file, but it is slightly unusual and can lead to confusion, particularly because a lot of documentation (mostly from third parties) is *horrifically* sloppy about "bytes" versus "words," to the point where you often can't tell what they are talking about.

There are a couple of immediate limitations we can get out of the way now: This PIC variant has only one 8-bit general-purpose register, referred to as W (the Working register). In times of yore, this register would have been called "A" for "accumulator"). Every byte of data you will ever process on the micro has to be funneled through this register. Also, the instruction set doesn't, strictly speaking, have any conditional jump instructions; it only has "skip next instruction if condition" instructions. While not exactly unprecedented—the AVR family has these instructions too, for example—this is a significant source of bugs from programmers who are not experienced with this style of programming, because the logic is backwards from the way most people are accustomed to think of binary decisions.

The PIC is a blindingly Harvard-architecture microcontroller, with *four* distinct address spaces: program memory, data memory and peripheral registers (Microchip refers to the on-chip RAM as the "register file," implying that it's really just a special case of on-chip registers), EEPROM and stack. More on the stack in just a moment; for now, let's look at the other two spaces. Program memory is simple enough; the reset vector is at 0000h, and the peripheral interrupt vector is at 0004h. The main code space stretches from 0008h to 03FFh. *Note that all those addresses are word addresses.* Each one refers to a 14-bit location, not a byte. This can be a little hard to come to grips with if you're used to working on larger architectures where, say, the middle of a 16-bit instruction at address 0000h can be accessed at address 0001h. You might be better off just pretending that each PIC instruction is a byte, in fact; words in the code space are atomic entities and can't directly be accessed by your software anyway.

The program counter in the PIC16F84A consists of two parts; the lower 8 bits are read/write-accessible in the PCL register, and the upper 5 bits are write-only through the PCLATH register. Simply writing to PCLATH doesn't immediately cause the CPU to start executing at a new spot—it's a kind of holding area (Program Counter LATcH, in fact) for the upper bits of the program counter. The CALL and GOTO opcodes only give you 11 bits of space for the destination address. If you're jumping to someplace far away, you need to save the W register, load W with the new desired value for PCLATH, store it in PCLATH, restore W, then CALL or GOTO wherever you wanted to go. If this sounds complicated, it is! The issue doesn't even directly affect the PIC16F84 (since it only has 1024 words of program space to begin with) but it is a real annoyance in larger parts of the same ilk.

The next oddity relates to the data memory space, which is 256 bytes in size and is banked into two 128-byte chunks on the PIC16F84A (there are more banks on the larger devices). The data memory bank currently in use is selected by means of two bits in the status register (RP0 and RP1; RP1 is not actually used for anything in the PIC16F84A and should be left zero at all times, although it is physically implemented in the core—some code uses this bit as a scratch flag). The following table shows the data memory map.

Bank 0	Name	Name	Bank 1
00h	INDF	INDF	80h
01h	TMR0	OPTION_REG	81h
02h	PCL	PCL	82h
03h	STATUS	STATUS	83h
04h	FSR	FSR	84h
05h	PORTA	TRISA	85h
06h	PORTB	TRISB	86h
07h	Not implemented	Not implemented	87h
08h	EEDATA	EECON1	88h

Bank 0	Name	Name	Bank 1
09h	EEADR	EECON2	89h
0Ah	PCLATH	PCLATH	8Ah
0Bh	INTCON	INTCON	8Bh
0Ch–4Fh	68 bytes of scratch RAM	Mirrors 0Ch–4Fh	8Ch–CFh

I've highlighted with a gray background those registers that are mirrored in both banks. I did *not* highlight the scratch RAM area, because it is merely a happy accident—due to incomplete address decoding—that this area happens to be accessible in both banks on the PIC16F84A. On PICs with larger RAM space, you cannot rely on this mirroring behavior.

Whenever you use immediate addressing (MOVWF and MOVF, which move a byte from W to a data memory location and vice versa, respectively), you specify exactly which "register file" address (to use Microchip's term) is the source or destination; the banking bits are ignored. The bank only becomes important when you use indirect addressing. On the PIC, indirect addressing is handled by the FSR (File Select Register) and INDF (INDex File) registers. You load a pointer into FSR, and this sets the target for indirect read/write operations. You actually access the indexed data through INDF, which is a dummy register; it refers to the RAM location wherever FSR, in conjunction with the register page bits, is pointing.

Implementing lookup tables in code memory is also significantly unusual in this family of microcontrollers. There is no instruction or addressing mode that permits you to read data directly out of code memory. Instead, the PIC16F84A offers an alternative version of the RETURN [from subroutine] instruction called *RETLW* (Return and Load W). This single-word instruction loads an 8-bit literal into the W register and returns to the caller. Any data table you need to have in ROM is thus actually a long string of RETLW instructions. In order to access the table, you have to calculate and load PCLATH, then calculate the low-order 8 bits of the address of the desired entry, and write this into PCL to perform a long jump into the data table. This is quite bizarre, not to mention terribly inefficient.

Next on the list of unusual behaviors, we come to the PIC's stack. I have almost nothing good to say about this system. In pretty nearly every other microprocessor architecture you will ever encounter, the stack pointer, if any, is a register that through some mechanism or other (occasionally quite convoluted), points into the device's general scratch RAM area. Often there are special rules about where the stack can reside and how large it can grow, usually caused by limits on the number of bits available for the pointer register, but nevertheless the actual data on the stack almost always reside in normal data memory.

The PIC16F84 is completely different from the norm—it has an 8-level x 13-bit hardware stack (13 bits are necessary because this is the size of the program counter), and the stack pointer is completely inaccessible to your code. All you can do is push and pop data using call and return instructions—interrupts also use the stack, of course. Since there is no exception thrown or flag set if you try to grow the stack below the hardware limit (it simply wraps around from location 8 to location 1, and there is no way to read the current stack pointer in software, it is impossible for your code to avoid, detect or trap stack overflow problems. The one "advantage" of this highly unusual system, if I can call it such, is that it is impossible for a runaway stack to blow away your data or program space, but this is something of a moot point if your program jumps off into hyperspace as a result of trying to pop a bogus return address.

While the larger PICs aren't *quite* as bizarre as the small parts, there is nevertheless a pervading feeling, even in the larger parts, that you're working on a really tiny micro that has had big RAM and Flash grafted onto it through horrific quasi-Frankensteinian operations. I can't help but think that somewhere in Microchip's design labs is an engineer who spent a few days too many packing ten-pound loads of product into five-pound bags, and went a little insane as a result. While it is perfectly possible to implement robust applications on a PIC, it's sufficiently annoying that I would greatly prefer to work on a different core unless there are significant factors (price, algorithm licensing, application-specific suitability or something else) that pull me toward a specific PIC device.

As I'm griping about the parts, let me raise another complaint—relatively few of the PIC family have an on-chip debugging interface. (Some of the very recent parts do have JTAG, and other parts support a sort of bond-out[19] debugging

---

19   Bond-out chips are special debugging versions of a microcontroller. They have additional pins brought out to external contacts. These pins are not normally led to the outside world in retail versions of the chip. Microchip's debugging interface operates on the same GPIO pins used for in-circuit programming; you can't use it if you're using those pins for other functions. However, the bond-out chip has extra physical pins that bring out the debugging lines to dedicated connection points.

feature using the MPLAB ICD 2 in-circuit debug module.) This means that usually the only available methods for code development are burn-and-pray or the purchase of a full in-circuit emulator. This might not be a huge burden to the home hobbyist (except for the expense of a full ICE), but it can be very irksome to use these parts in commercial applications where development time is money. Some applications—particularly radio devices, which is where I work at the moment—are very difficult to debug when you have to attach an enormous emulator spewing RF interference. JTAG is bad enough, but it's easier to mitigate this noise somewhat. Your ICE, on the other hand, needs to be physically close to the board to avoid undesirable parasitic effects from long cables stretched from the ICE into the location where the microcontroller should be.

From the development hardware perspective, there are numerous third-party PIC burners on the market, ranging from simple homebrew devices connected to a parallel or serial port to high-end multigang programmers. (PonyProg, which I mentioned earlier, also supports several PIC variants.) The entry-level burner from Microchip is the PICstart Plus. This is a small, inexpensive, serial-connected device with a DIP ZIF socket on it; you can burn most of the popular PIC variants with this hardware, though you'll need to make a homebrew adapter to do in-circuit programming. Microchip also makes several other burners of various capabilities, as well as full-speed in-circuit emulators like the ICE2000 and ICE4000. Third parties such as Vikon *<http://www.vikon.com/>* also make ICEs, slightly cheaper than Microchip's product.

On the development software side, the PIC is possibly the least high-level-language-friendly architecture ever devised. This range of parts is best programmed in assembly language, particularly if you're tight for code space or looking for optimum real-time performance. However, commercial PIC C compilers are available from Byte Craft Limited *<http://www.bytecraft.com>*, CCS Inc. *<http://www.ccsinfo.com/>*, Hi-Tech Software *<http://www.htsoft.com/>* among others. Hi-Tech also has a freeware product, PICC Lite, which supports ten commonly used hobbyist PICs with some limitations (depending on which device you're using, there may be either no limits, or RAM/ROM usage limits).

For free tools, your best jumping-off point is *<http://www.gnupic.org/>*—this site links to numerous ongoing open-source toolchain development projects. You'll find several C compiler projects (of which two are at a usable state at the moment) as well as three Forth compilers, a Python compiler and more. There is even a Java-to-PIC-assembly translator, called *Aino*, available from *<http://personal.eunet.fi/pp/jokinen/>*. While I was discussing this latter product with a nonnative English-speaking colleague, he made the comment that "tattoos do not make the pig." This statement is cryptic and slightly bizarre, and completely summarizes my reaction to the idea of Java code inside a PIC.

All this leads me to summarize that (strictly in my personal opinion, of course), the PIC family is best suited for relatively small projects, preferably in assembly language.[20] While I'm aware that some quite amazing things have been done in PICs, I would say these efforts are misspent unless there are some very special circumstances that make a PIC the right choice—price, some very specific combination of features, and so on. While the PIC family is not *terrible* for educational purposes, there are other options that are easier for a beginner to deal with. Hence, I wouldn't recommend this family as your starting point into embedded engineering, despite its apparent ubiquity.

Oddly enough, there are clones of a few PIC products—the "big name" in this field is Ubicom™, formerly Scenix. Every now and then, a small Far East company will come out with a few parts that are suspiciously similar to some PIC variant, but these vendors always seem to fade out of existence in fairly short order.

3.6   Less Common Architectures for Special Needs

As you'll appreciate, the sections preceding this one have barely scratched the surface of enumerating what sorts of 8-bit (and smaller) architectures are available. In no particular order, here are a few other architectures that you might encounter in real life:

- DSPs of all flavors.

- Rabbit Semiconductor®'s high-end "kinda Z80" processors.

---

[20]   This comment doesn't apply to a few very special-purpose parts like the dsPIC family and the rfPIC devices.

- 68HC09 from Freescale™, nee Motorola.

- Embedded Z80 variants from Zilog®.

- COP® from National Semiconductor®.

- NEC's 78K series.

- Special-purpose parts for speaking toys and low-end LCD or video games (many of these are based on variants of the 65C02 or 65C816[21] core).

There are various special considerations and heuristics that might lead you immediately to select or reject a particular microcontroller family. Since all of these rules represent application-specific factors, it's more or less impossible to present them as a coherent Grand Theory of Microcontroller Selection; it's simply a set of questions you have to ask yourself, then weigh the answers to reach a conclusion.

— *Have you used this architecture before?* It seems obvious, but people all too often under-weight this consideration. If you already have experience working with the basic architecture, you should give strong preference to leveraging that experience.

— *Is the published technical information enough for you to estimate accurately if the part actually meets your needs?* (If not, can pre-sales support fill in the gaps for you?) It is a very unpleasant thing to discover, eight months into a project, that the microcontroller you selected isn't *quite* fast enough, or generates just a little too much RF noise for your application.

— *Are there drop-in variants of the part you're considering with more resources (ROM, RAM, faster clock, and so on)?* With the best will in the world, you can't always estimate your software requirements accurately—plus, marketing is always pleading with you to add just one more feature. It's a nice safety net to know that if you run out of RAM or code space, you can simply upgrade to a bigger, pin-compatible version of the chip.

---

[21] The 65C816 is a backwards-compatible, 16-bit version of the 6502. It was used in the Super Nintendo video game, the Apple IIgs, and a few other pieces of home computer and/or video gaming hardware.

— *Are other people using this family in varied applications?* Some microcontrollers are designed specifically for a particular narrow application (say, CD player servo control). You might look at one of these chips and think it's an ideal fit for your new laser-guided golf club widget—and it's cheap, too, because they're making ten million CD players a day in China. Problems will, however, arise when the primary users of the chip migrate to something else due to technology changes in their industry—volumes from the secondary general-purpose users like yourself will probably not be sufficient for the manufacturer to continue production. At the very least, you can expect the family size to be reduced as unpopular parts are shed; prices will likely jump, too.

— *Has this device been generally available for a number of years?* Somebody has to be on the bleeding edge, trying out new parts on the day they're released. Unless you're paid to research these things, you don't want to be that somebody.

— *Are there multiple sources for development tools?* Single-source development tools are a red flag that the chip vendor had to contract out development of the toolchain, or (which is often worse) the sole tool support may have been developed in-house. This is rarely a positive sign unless the part you're looking at is really, really new. And if it *is* that new, shame on you for thinking about using this part unless there is literally nothing else on the market that will do the job.

— *Is this a Far East part intended for their domestic market?* All too frequently, Far East silicon vendors will look at their lineup of domestic-sales-only devices and decide that it's time to start selling some of these parts in the United States. They'll use some arbitrary process to decide which specific parts should be sent across the sea, and all of a sudden your local rep will call trying to sell you some "brand new" chips. I've been burned by this sort of thing on more than one occasion. The English documentation and development tools are often barely ready for primetime, and your local field applications engineer will most likely know nothing about these devices.

Note that I'm not making these complaints about Far East parts in general; I'm very specifically referring here to parts that have historically been available only to domestic customers in Japan and China. It's hard to guess at the reasons these parts are so abominably poorly supported in the United States,[22] but my surmise is that the engineers who are using these parts in the geographical region where they originate have much better contact with the chip designers. Hence, there's a lot of tribal knowledge that is communicated orally between the semi manufacturer and their domestic customers, and this information simply never gets written into the datasheets or errata, and certainly is never translated into English. As for the development tools, the most charitable explanation I have is that the English versions might not be kept as up-to-date as the Japanese/Chinese versions.

— *Is this a special-purpose part* explicitly *made for carrying out your application at minimum cost?* The microcontrollers used in toys, with which I have some intimate experience, are rather unique; they generally have very limited capabilities, and toolchain support that is little short of shocking. As you might guess, the instruction set of these chips is not very rich. They typically have instructions to load a register with a constant, increment/ decrement a register, set the state of the output pins, jump unconditional, jump unconditional through a register, and make various conditional jumps based on input pin state. These chips, at least at the low end, often have no RAM at all except for a few registers. There is no stack and the only way to implement a subroutine is to store a return address in a register. If you're programming these devices, ignore everything your computer science professor taught you; the only way to write a program is to use a forest of GOTOs and, for any nontrivial project, usually a lot of almost-redundant code. However, they still represent the best choice for the application, because toys typically don't require much in the way of complex logic (hence, you won't be spending much time programming and debugging), and toys are extraordinarily price-sensitive.

— *Is this part a proprietary core from a small, fabless vendor?* While it's not an instant disqualifier, it should be a yellow flag to you if a device is both

---

22   European readers should note that I'm being very specific about the U.S. market here. The comments I'm making in this paragraph do not always generalize well to the European markets; some distributors there provide very good support.

completely proprietary and from a small fabless company. If the design company has problems with their fabrication partner, you'll have supply issues.[23] This factor needs careful attention if you're designing a product for consumer-scale mass-production, particularly over the long term (HVAC equipment, for instance, might fall into this category). If you're building something for small-scale or short-term production, you will probably pay more attention to convenience; how easily you can get a part that contains all the functionality you need, preferably with pre-built driver libraries. Rabbit Semiconductor's products are an example of a family that you probably wouldn't choose for a high-volume application. They're well-supported for rapid development, and they offer a package of features that would require significant engineering effort on your part to assemble around a different micro. However, intellectual property reusability and scalability across a broad product range is hard to achieve with these parts (the family is very small and the primary development environment is basically proprietary). The chips are also relatively expensive.

— *Is the core sole-sourced, or is it licensed to multiple vendors?* Not many micro-controllers have drop-in second-source replacements in this day and age. Even in corporations that have strict rules about not specifying sole-source parts, the rules are usually waived for microcontrollers. However, even if you're not looking for an *exact* replacement, there is some utility in choosing a core that is available from multiple vendors. If you have a falling-out with vendor A, or they discontinue your pet part, and the same core is available from vendor B, you will probably have less difficulty designing out vendor A than if you need to do a complete port to a new, unknown core.

Hopefully by this point, you have a reasonably good idea of why there is no single microcontroller that you MUST learn in order to get into the embedded field. You should also have some idea of the pros and cons that go into selecting a part for a real project, and I've also done my best to give you a quick

---

[23] A fairly recent example of this is the debacle surrounding ZF Micro Solution's parts. These are x86-compatible system-on-chip devices formerly made at National Semiconductor's fabs. There are several stories about why this relationship went bad, but the net result is that all of ZF's customers were out in the cold for quite some time until a new fabrication deal was inked. When you design in a specialty part like this, you tie your fortunes to those of your supplier—like it or not.

thumbnail of the features and oddities of a few of the most popular learning platforms. If you're just getting started with embedded programming, your best path to employability is to take *any* one of these architectures (or another part that captures your interest) and start building real projects with it. As long as you meet the design requirements of the project,[24] the particular chip you use to build it is unimportant.

3.7   What Programming Languages Should I Learn? C++ vs. C vs. Assembly Language in Small Embedded Systems

I'd like to open this section with an illustrative piece of text that I originally posted to the Usenet group comp.arch.embedded (c.a.e) early in 2006. I've edited it just a little; you can find the original in Usenet archives if you're curious. Note, for the record, that this posting is very much tongue-in-cheek on my part, and I'm quoting it here with the same spirit.

(Original poster, a fellow Elsevier author in fact: "Most everyone uses C or C++ these days."[25])

I take a bit of exception to the idea that "everyone uses C++," especially in the embedded arena. I guess it's been a few months since this particular chestnut was sizzled on the coals in c.a.e., so maybe we should just cut and paste the last argument thread to save everyone's time.

I'm starting to see an approach that approximates sanity with regards to the use of Java in high-end embedded projects, rather than C++ (a language which is abhorred on any architecture by all right-thinking embedded engineers—and of course, only an exceptionally jaded masochist who has lost the phone number of his dungeon mistress or can no longer afford her fees would contemplate C++ on an 8-bit micro).

If you're prepared to live with the overhead of C++, and you're convinced that there is something to be gained in your application from a language that has OO

---

[24]   In the real world, this would include price and time-to-market, of course.

[25]   Please note that I was *intentionally* subjecting the original poster to selective quoting in order to construct a humorous response.

syntactic elements (like C++) rather than merely OO capability (like C), using Java instead can be a pleasant surprise. Among other advantages of Java over C++, it is occasionally possible to find someone who can answer a Java question both exactly and correctly. Furthermore, positively several Java questions do not have as their "real" answer a 5,000 word argument among five language lawyers. No C++ question can be answered without either:

a) employing several language lawyers of opposing viewpoints, and/or

b) developing a new draft standard defining a version of the language with the desired functionality.

Dogma is a wonderful thing, isn't it?

Commerce degrees used to include a subject called *creative accounting*. The core of this subject is: Your company bought $X, sold $Y and received $Z over the past year. Your CEO, stockbroker or anonymous penpal wants the company to show a profit of $F. Obtain this result in a GAAP[26]-compliant way.

In the past, I've seen vast practical use for this course work in terms of justifying (or killing) value engineering projects, obfuscating projected bill-of-materials costs, and demonstrating that a department has or has not met its budgetary goals. But the single most useful fact it taught me is that all metrics not directly derived from underlying measurable physical phenomena are nonsense—and cost/productivity/reliability savings realized by moving to C++ are at the extreme bad end of this spectrum. Moving from assembly to C at a certain [fuzzy] size/complexity point can be justified merely in terms of the number of people that are comfortable programming in the language (though I am quite convinced that there's more gain than merely this one point, and most c.a.e readers would agree with this).

Moving to C++ in embedded projects* requires justifications that typically end in the words "MY GOD! A MAUVE DOLPHIN ON A BICYCLE IS FLYING PAST THE WINDOW!" (Alt-F4 your PowerPoint drivel while everyone looks out the window.) "This concludes the presentation; in summary, you should do what I say."

---

* Special exceptions apply. These are mostly ludicrous; for example, designing a project that is not a PDA around Windows CE.

---

[26] Generally Accepted Accounting Principles. You can't break the rules until you know what they are.

Engineers are a contentious lot, as the preceding snippet should amply demonstrate. The issue of which programming language represents the One True Faith of embedded software development has been the cause of literally millions of words of finely reasoned technical arguments sprinkled with sample sourcecode, and not infrequently interspersed with bitter wrangling. I'll give you a sneak peek at the truth right here in this first paragraph by stating up-front that there is, in fact, no One True Faith, and that the correct programming language to use for an embedded application is a situation-specific decision. (Naturally, there are plenty of people who would argue this statement, just as there are people who would defend it to the death. In fact, merely by omitting [insert name of any computer language here] from the title of this section, I've invited flames down on my head.)

The decision as to which programming language to use for a project rests on the following legs, in no particular order:

- The complexity of the project, which usually translates more or less directly into code size.

- The project manager's preferences, experience and, unfortunately, arbitrary whim.

- Company policies.

- The availability within your organization of experienced programmers for a given language or tool. Though this is clearly not universal advice, it has been my experience that outsourced programmers have a predilection for high-level languages even when other aspects of the project are decidedly in favor of assembly language.

- Availability of trusted tools for the hardware platform in question.

- The resources (ROM, RAM, CPU horsepower) available in the hardware platform. As resources become tighter and tighter, hand-tweaked assembly language starts to become the only viable option.

- Functional requirements of the system; in particular, real-time responsiveness. The further your language abstracts you from the hardware, the more difficult it is to guarantee overall system real-time behavior.

- Reliability and safety requirements enforced by a regulatory body (in particular, certifications that might be required, and documentation that may have to be generated in order to prove that a system exhibits, by design, a given level of reliability). For example, it might be necessary to show simulation results to prove system robustness.

- The embedded operating system, if any, in use. This point can often be the force driving all other considerations. If your application has relatively complex requirements (for example, support for exotic networking protocols), vast engineering time savings can be realized by using an off-the-shelf operating system with ready-rolled support for the functionality you require. You are then, however, usually constrained to use the preferred language for the operating system.

Most of us are not designing spacecraft, defibrillators or automobile airbag control systems. People who do work on such systems have many design decisions made for them up-front, in the form of regulatory agencies and company policymakers publishing a list of commandments stating "thou shalt use toolchain X and code style ABC." At the risk of generalizing very unfairly, I'd say that the vast majority of embedded engineers work at the other end of the spectrum, on projects that have few or no regulatory requirements and very little in the way of safety implications.[27] Therefore, it is necessary for the engineer to be able to select a toolchain intelligently, or at least understand the issues involved.

Most of my midsize to large projects, just like the majority of other embedded projects in the world, are written in a mixture of assembly language and C. My small projects are generally written in pure assembly language. As an *extremely* rough rule of thumb, I tend to see the cutoff point between "assembly only" and "maybe go to C" as lying somewhere between 4K and 8K of object code. This rule is so heavily laced with caveats, however, that I'm very reluctant to offer it.

The space that C++ should occupy in the embedded systems arena is a fiercely debatable topic, as the start of this chapter probably demonstrated to you. My

---

[27] By this I don't mean that most systems wilfully neglect safety considerations; I mean that most of these systems are of such a nature that the most catastrophic programming failure imaginable has no conceivable safety consequences. If your Furby doesn't speak to you one morning, there are unlikely to be serious, life- or property-threatening repercussions.

own view (and this time I'm not speaking tongue in cheek) is that C++ has no place at all in 8-bit development. Despite what computer scientists would have us believe, the object-oriented programming approach does not lead to significant or even merely measurable productivity, maintainability or reliability gains in the closed environment of a low-end embedded system. This is true, in my opinion, even when the object-oriented features of the language are employed consistently and correctly, which is almost never. Others clearly share my point of view, as this quotation illustrates.

[. . .] reuse and integration of independently developed embedded C++ components is difficult for a variety of reasons. First, the C++ standard makes specific provisions for implementation-dependent and implementation-defined language features where an integer might be 32 bits on one processor and 64 bits on another. Second, the C++ standard does not fully specify the behavior of the standard template libraries (STLs) or underlying operating-system services, nor does it standardize the collections of services needed from the operating system. Third, both the C++ language and the STLs are large, complex, and quite difficult to understand. Fourth, C++ does not offer automatic garbage collection. If the integration of components shares memory references between components, the system integrator must determine when those shared objects can be discarded and which component must reclaim memory.

A typical C++ component has buried within it hundreds of implicit dependencies on the target processor, C++ compiler, host operating system, and STLs. Assumptions are rarely documented, and most C++ programmers do not realize that the validity of their code depends on assumptions not necessarily valid in all execution environments. The typical C++ programmer simply tests and refines code until it works in the test environment.

—Kevin Nilsen, "Java Sounds the Death Knell for C++," *Electronic Design*, May 26, 2005. Original article available on the Web at *<http://www.elecdesign.com/Articles/ArticleID/10274/10274.html>*.

As a result, if you ask the question "which languages should I learn?", I would say that it is essential to learn C really thoroughly, and at least one assembly language to a reasonable level of competence—in the 8-bit field, it's advisable to get as good as you can with assembly language, even though in this day and age a lot of your code may be written in C. In these sorts of systems, you're almost always going to need at least the skill to read your compiler's assembly output and understand what it's doing. Compiler bugs are more frequent in these low-volume compilers; you need to be able to detect them, report them and work around them—but even more important, you need to be able to hand-optimize your code, where necessary. You might also have to write your own startup code, or modify the compiler-provided startup code, for some applications; it's really not possible to get away with absolutely zero knowledge of assembly language (despite what compiler vendors may try to tell you!).

While it's useful to have some familiarity with the assembly languages of several different microcontrollers (simply so that you know some of the different ways in which particular things can be done on different parts), it isn't usually a go/no-go issue as far as employment goes. Certainly, I wouldn't worry about listing 50 different microcontrollers on your résumé. The reason for this is simply that once you have experience with a few different architectures, learning the assembly language for a new processor is just a couple of days' work with the datasheet, unless perhaps the part in question is radically unusual and/or not adapted for programming by hand (such as a VLIW architecture). If you list four dozen different cores on your résumé, employers are likely to think that you just gathered a list of everything you could think of and stuck it on there. Confine your résumé to discussing parts on which you've actually carried out major projects; prospective employers want to see achievements, not waffle.

3.8   Brief Ravings on Copy-Protected Development Tools

The issue of whether it is generally preferable to use closed source commercial development tools or open source tools is an exceedingly knotty one, and the subject of numerous arguments. Since I hopefully have you glued to the page and hanging on my every word by this point, I'm going to take the opportunity to instill into you the doctrines of the One True Faith.

The underlying premise of what I'm about to say is as follows: Code in embedded systems tends to last a very long time. Once an investment is made in a particular microcontroller and/or a certain internal software architecture, changes from that point tend to be incremental, and in many cases driven only by explicit customer requests. Moreover, embedded systems are mostly self-contained, sealed code ecosystems. In the vast majority of cases, your user will never add code to your system, upgrade the operating system or interact with the microprocessor in any way except via the inputs and outputs you provided in your design. Furthermore, many embedded systems are devices that have an indefinite lifespan; they do not have a perceived "coolness" factor that, when dissipated, makes the device worthless. (Contrast a computer game—which your kids will throw away once they want the next great thing—against the powertrain control module computer in your car, which will go on doing its prosaic thing until it meets the crusher.)

One consequence of these facts is that embedded code can lie dormant for a long time between initial release and subsequent revision. Product lifecycles are typically in the neighborhood of two years in most of the consumer electronics arena. So you may release a product, tuck away the sourcecode and development software and hardware into a bottom drawer, and move on—only to have marketing come back to you a couple of years later saying they want some revisions so that the product will better fit into this year's lineup of widgets.

If you're cursed with copy-protected development tools, this is usually the point at which you find that the new PC you were issued last year doesn't have the right sort of ports to talk to the hardware dongle that came with your old compiler. Or perhaps the new operating system you were forced to accept runs the compiler software in a virtual-machine sandbox where it can't access the hardware enough to fulfil its copy protection requirements. Maybe you simply need a new magic authorization code to reinstall the software on your new computer, or maybe the dongle was eaten by your dog. Whatever the case, this is the point at which you call the compiler vendor and they either don't exist (in which case, you have the wonderful choice of either heading off to an Eastern European cracks'n'virus emporium for a crack, or buying a totally different toolchain), or they say that the version you're running is too old for the modern day (shame on you!). By the way, of course, your free support period has expired.

In these situations, you'll typically be required to buy the latest version of the compiler, because of course the vendor can't be expected to fix all the copy-protection-related bugs in old compilers. So, not only will you be losing a significant wad of cash, but you'll also have the joy of porting your code into a new and untried compiler version. The first step when recompiling old code is normally to compile it unmodified and compare the binary with what you wound up with when you built it years ago, in order to make sure you have all the settings, paths and files correctly configured. This kind of sanity check is going to be impossible with a new compiler, so you won't even know if you have successfully replicated the build environment. Extra overtime all round for all of engineering and QA—and all because of an explicit act on the part of the compiler vendor.

I see this kind of thing happening around me all the time, and I have difficulty comprehending why more people don't complain about it more loudly. If I buy a license for XYZ software, and the license doesn't EXPLICITLY say that my right to use the software ends in a year,[28] then I should be able to run the software for as long as I have fingers with which to type and a machine that is capable of running the program. I should *not* be barred from using the software just because of a failure or obsolescence of an otherwise useless piece of gatekeeper hardware or software.

In order from most to least hateful, here are the types of copy-protection and license schemes you will see on embedded development software:

- Hardware-based copy protection schemes; dongles, mostly. It's particularly annoying when these sorts of amazingly intrusive and trouble-prone copy protection systems are applied to software that requires special hardware to run anyway.

- Live authentication schemes that require a real-time communication link between your PC and the software vendor's home base every time you run the program (and often periodically while you're using it).

- Challenge-response schemes where the software generates a new, random challenge every time you install it—you need to call or email the vendor for a new response code every time the software is reinstalled, even if it's on the same machine.

---

[28] Software like this—restricted by an annual service contract—does exist, of course. Run screaming if someone tries to rope you into a deal of this sort.

- Challenge-response installation schemes where the challenge is fixed, based on some aspect of your hardware. If you reinstall the software on the same machine, you will not need to contact the vendor; you only need a new key if your hardware changes.

- "Branded" software that requires a key file containing your name and other details in encrypted form. Any copies made of this software will operate correctly, but will show your contact details onscreen so everyone knows where the software was copied.

- Closed-source, but unprotected software—can be installed anywhere at any time.

- Open-source software, where you get to choose when and where you install the program. If necessary, you can rebuild it under a new operating system.

If you read any of the previous sections in this chapter at all, you'll know I am a strong advocate of open-source tools. Tying up intellectual property inside software that you might not be able to run on your next PC just doesn't make sense. Open-source tools are ideal because not only can you archive them (and give a copy to your customer, if necessary), but you can rebuild them on new and alien operating systems in the future.

Of course, in some cases, open-source just won't do the job. Sometimes you have to use existing code libraries in proprietary formats, and sometimes you need a special optimization capability found only in a commercial product. The free compilers are not supported by a raft of cash and free device samples, so they don't have the same update priorities as commercial products (though this doesn't always mean that they have poorer performance).

When you get out in the workforce and need to select development tools, you'll read a lot of argument on both sides of this issue. I strongly suggest that you think forward to the time when you're going to need to support legacy code built with an ancient compiler, and select open-source tools by preference. Vendors who use outrageous copy-protection technologies on their tools are playing games with your future profitability.

3.9 An Example 8-Bit Project Using AVR and Free Tools

In this section, I'm going to illustrate a relatively small educational AVR project. Much of the text in this section was first published on IBM's developerWorks site. You can see links to the original articles, along with downloadable source-code, at *<http://www.larwe.com/technical.html>*. I chose this project because it demonstrates a lot of the thinking you'll have to do when building portfolio projects of your own.

I started by defining, at a block level, what the overall system is supposed to do. This design is for a robotic submarine project called *E-2*, based around a PowerPC network attached storage (NAS) appliance called a *Kuro Box*. The Kuro Box is a Japanese invention; essentially, it's a little PowerPC single-board computer running Linux.

It was originally sold as a NAS device under the name LinkStation; however, it proved so popular for hacking purposes that the manufacturer now sells the Kuro Box (essentially a LinkStation without a hard drive in the box) as a separate product. The name means "expert box," reflecting the hacker target demographic.

Kuro is based around a 200 MHz MPC8241 (PPC603e core). It has 4 MB of linear boot Flash, 64 MB SDRAM, 10/100 wired Ethernet, a USB 2.0 port (host-side), and an IDE interface. It was available locally in the United States from Revolution, *<http://www.revogear.com/>*, for $160, but has now been superseded by the $149 Kuro Box HG WR. This price buys you something that approximates a turnkey system. You merely have to install a standard 3.5 inch IDE hard disk and run the vendor-supplied Windows® setup utility, which partitions, formats and loads the drive contents over a telnet connection.

Since you're probably not familiar with this device, Figure 3.5 shows a block diagram of the Kuro Box. Note that it does not show all the MPC8241's peripherals; it only shows those parts of the chip that are involved in using the interfaces I used for this text.

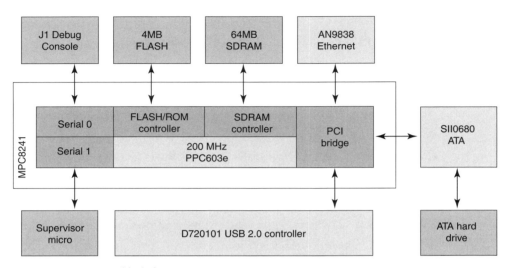

Figure 3.5 Kuro Box block diagram.

The MPC8241 microcontroller provides (among other features) an on-chip PCI bridge, SDRAM and Flash controllers, and two serial ports. The first serial port is connected to J1, the debug console port. The signal swing here is from 0V to +3.3V, and it is inverted for compatibility with RS-232 level shifters. The second serial port is connected to the slave microcontroller (an AT90S2313). This micro handles power sequencing and fan tachometer feedback, and also provides a master watchdog for the MPC8241. The software bundle preloaded on the Kuro Box includes a daemon that kicks this watchdog periodically. CN1 on the motherboard is a standard 6-pin Atmel AVR ISP port connected to the slave microcontroller.

The board also offers a normal COP/JTAG debug port, although the connector is not populated as shipped from Buffalo. In order to use this port, you should add the 10K series VIO resistor R67 and a 4-way 1K resistor pack at RA11. The least expensive route to using the COP port is through a "wiggler" connected to your PC's parallel port; the cheapest commercially available product for accessing the MPC8241's on-chip debugging facilities is the Macraigor Wiggler.

In this text, I'm going to use a lot of acronyms specific to the E-2 project. I'm documenting these on my personal website at *<http://www.larwe.com/sub/glossary.html>*, but the first two acronyms I'd like you to remember are VCM (Vehicle Control Module) and SCM (Science Control Module). In the context of this text, the SCM is the Kuro Box, and the VCM is the small real-time board I'm describing in this chapter. Figure 3.6 shows a block diagram of the system as a whole. The VCM is essentially everything that's not inside the Kuro Box, excluding the power systems.

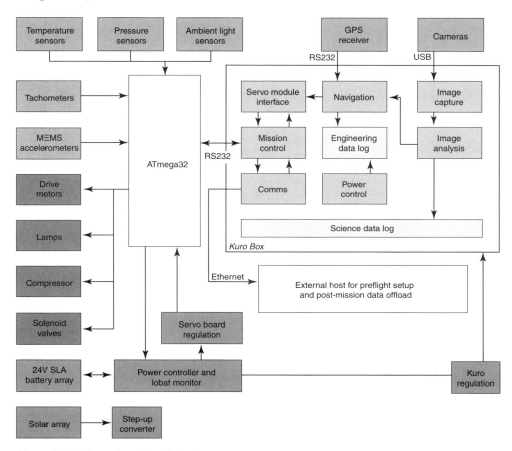

Figure 3.6 Robot submarine block diagram.

The screen coding in this image is as follows: ☐ storage ▦ software component ▦ power system component ■ actuator/output component ▦ sensor/input component.

In the special version I built for developerWorks, I'm using a 7Ah battery pack (specifically, two Powersonic PS-1270 12V/7Ah batteries) and a solar panel with nominally 36W of output. This is a convenient test setup that I have on my workbench at home; it's essentially a scale model of the power systems on the actual E-2 vehicle. I find it inconvenient to have hundreds of pounds of lead-acid batteries sitting next to my lab bench!

For those who are really interested, here are some more details on how this technology is applied in my submarine project. If you're not building a vehicle of your own, you can skip the next few paragraphs. Although the hardware you see here is very similar to the control system in the sub, there's an important architectural difference. In the sub, the heavy processing iron (currently an Advantech PCM-5820 single-board x86 computer based on the AMD Geode™ processor) doesn't do the navigation tasks; it is powered down almost all of the time in order to conserve energy. In fact, for the daylight hours, the submarine is configured in a minimum-power mode with practically everything switched off, so that the batteries can reach the maximum possible charge level. The actual mission is carried out at night—the rationale for this being that underwater you'll need lights to see anything interesting regardless of what time it is, so you may as well spend the daytime just collecting energy.

As a result, the block diagram of the E-2 is slightly different from what you see in this text; it has all the same blocks, but the GPS receiver goes directly into the ATmega32. Through a little software and hardware cunning, a single UART on the ATmega32 services both the GPS and the SCM. From a conceptual standpoint, the other main differences are that all the software feedback loops (depth control, for instance) are closed within the ATmega32, and the engineering data log (EDL) is connected directly to that micro as well. The EDL is implemented on a standard SD or MMC card in SPI mode.

This architectural decision, by the way, is why I don't need to worry very much about security on the Linux side of the equation (the Kuro Box runs a standard edition of Hard Hat Linux without special security extensions). During normal operation, the outside world simply cannot communicate with the Linux box. The only field communications interface during mission time is with the VCM, which is as secure as it needs to be.

Note that there are also a couple of additional function blocks in the E-2, designed to communicate with the device and help with recovering it if lost, which this article series doesn't get into. These modules will eventually be described in detail in the E-2 public information area on larwe.com; one of them is an emergency recovery beacon similar to an aircraft flight data recorder's "pinger," and the other is an off-the-shelf satellite transceiver. E-2 can be contacted and commanded using satellite Short Message Service (SMS) strings; telemetry data can be uplinked over a moderate-bandwidth commercial satellite phone subsystem.

Enough digression; back to the task at hand. First, let's get acquainted with a few of the design requirements for the board: Many of the actuators (solenoid valves, motors, and so on) needed in this project are most readily available in 24V flavors, so the board should be designed to operate off a 24V battery supply. This raises the first interesting question of how to regulate the bus voltage down to a micro-friendly 5V.

Being a lazy sort, I'd usually take the easy way out and dump a simple linear regulator into an application like this, but dropping 24V down to 5V is just too big a step, as it would mean throwing away about 5W in that regulator for a relatively modest 250mA load. Use of a switcher is therefore mandatory, and if you refer to the schematic you'll see that I've used the dirt-cheap MC34063A in a direct crib of its step-down application note. There is nothing at all special about this circuit, and if you're trying to throw together a quick hack to experiment with the code I'm talking about here, feel free to use a 7805, or just power your board off a lab power supply. Substantial heatsinking and a TO-3 package on the regulator is necessary if you use a 7805 with a 24V input voltage.

Fulfilling Step 3 in the goals previously mentioned requires choosing a sensible hardware interface, and overlying software protocol, between the VCM and SCM. There are many ways to achieve this, but the method I've chosen is an asynchronous serial interface with RS-232-compatible levels. Please note that this probably wouldn't be the ideal choice for a real integrated application that you were building from scratch. If you were using a bare MPC8241 part, for instance, the right thing to do would be simply to put some 3.3V-5V level shifting in place and connect directly to one of the MPC8241's serial ports. Unfortunately, we're fighting the Kuro design a bit here; one serial port is reserved for debug output, and the other is connected to the supervisor/watchdog chip.

The great thing about async serial connections, however, is that they can be plugged into almost any hardware platform on the planet. In the case of the Kuro Box, you can either go in through the debugging serial port, or you can put a standard level-shifter on the micro and attach it to the Kuro Box through a USB-to-serial converter.

This discussion follows the latter route, partly because the debug port is constantly in use by the kernel, but mostly because this way you can use the same VCM hardware to talk to a Microsoft® Windows® machine running Cygwin, or a regular PC running Linux, or indeed almost anything else with a standard serial port.

I'm using a Keyspan USA-19W USB-to-serial adapter in this project, because I happen to have it lying around; there are numerous other such adapters supported by Linux and the standard kernel has modules for the Keyspan adapters, among others (Belkin adapters as well as FTDI USB-serial solutions). The only thing you might need to do is to create the necessary /dev/ttyUSB* device nodes.

By the way, the debug serial port, J1, is an essential route into the Kuro Box, since it is the default console for kernel messages and bootloader communications, and it is the easiest device to use for kernel bring-up and debug. For inscrutable reasons, Buffalo neither documents nor fully implements this port. The pinout of J1 is as follows (the pads are drilled for a standard 100 mil SIL header, which is rather a relief as most of the other connectors in the appliance are on 2 mm pitch):

Pin	Function
1	TxD (data out of the Kuro Box)
2	RxD (data into the Kuro Box)
3	3.3V power out of the Kuro Box
4	Ground

As is fairly standard with debug ports on embedded boards, the serial port is NOT at RS-232C levels; it is driven at the CPU's I/O ring voltage (3.3V) and it is inverted relative to RS-232C. This means that you need a charge-pump type inverting serial transceiver IC in order to interface it to a normal PC. In turn, this usually means breadboarding something quick and dirty.

However, there is an easier route, which also gives you a one-stop solution to the problem of having a legacy-free PC. Most USB-to-serial converters consist of a microcontroller that does the USB interface, and a separate, off-the-shelf transceiver chip doing the level translation. As a rule the I/O voltage in these devices is 3.3V, which is conveniently compatible with the MPC8241's I/O ring.

I chose to use a PalmConnect® USB adapter, originally intended to adapt a Palm III™ or m100™ cradle to USB. Figure 3.7 shows the component side of the PCB of this adapter.

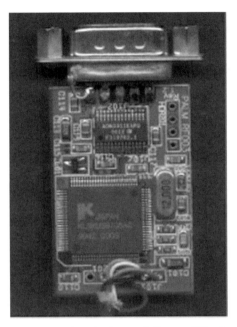

Figure 3.7 Serial adapter, before modification.

If you have a different adapter, you'll certainly see a different layout and different micro, but it is likely that the transceiver part is something very similar to the Analog Devices ADM3311 used in the Palm® adapter, if not the exact same device.

The first step you should take after opening your serial adapter is to check that it uses 3.3V logic levels internally. (If it uses 5V internally, the current-limiting resistors and 5V-tolerant I/Os in the Kuro Box will protect it, but you might not see any output on the PC side.) To determine the internal logic voltage, simply plug the adapter into a USB port, wait for it to be recognized by the host PC, then use a multimeter or oscilloscope to probe the supply pins on the transceiver chip—pins 23 (GND) and 2 (Vcc) in the case of the ADM3311.

Now you need to desolder that transceiver chip. The method I use is to put a bead of solder down all the pins on one side, then get an Exacto blade under the chip to lever up that side while keeping the entire bead melted with my iron.

Repeat on the other side. You'll very rapidly seesaw the chip off the board this way; just be careful not to break any traces. Practice on some junked appliance if necessary. Figure 3.8 shows the board after desoldering the transceiver.

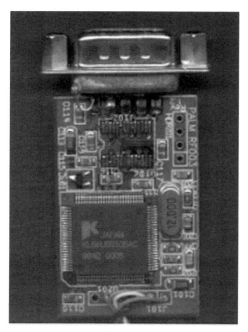

Figure 3.8 Serial adapter, after removing chip.

Now you need to determine where to feed the Kuro Box signals. To do this, you must establish which gates in the transceiver handle RxD and TxD, and connect to the "other" side of those gates. This involves tracing back pins from the DB9 to pads of the IC you just desoldered, then referring to the datasheet for the transceiver chip to work out where the other side of that gate lies.

Pin 3 of the DB9 is TxD, and the prior example runs to pin 22 (T1OUT) of the transceiver. The corresponding input pin (T1IN) on the transceiver is pin 7, so we run a wire from pin 2 (RxD) of the Kuro Box's J1 to pin 7 of the space where the transceiver used to be. Remember that we're effectively implementing a nullmodem here, which is why we route TxD to RxD and vice versa.

Similarly, pin 2 of the DB9 is RxD, and in the previous example runs to pin 19 (R1IN) of the ADM3311. The corresponding output pin (R1OUT) is 10, so we run a wire from pin 1 (TxD) of the Kuro Box debug port to pad 10 of the transceiver's space.

Finally, we connect pin 4 (GND) of the Kuro Box debug port to any convenient ground point on the USB-to-serial adapter. The 3.3V supply line from the Kuro box isn't needed, so don't connect it to anything.

Figure 3.9 shows a suggested way of routing the cable: Connect it to the solder side of the Kuro mainboard, use regular 50 mil ribbon cable, and route the cable out of the fan exhaust hole. Obviously, do not route it through the fan itself! There is a gap between the fan and rear bezel; it is ample space for the cable to exit without touching the fan blades.

Figure 3.9 Serial cable routing.

Figure 3.10 is a close-up of the patched PalmConnect board. Tweezers are necessary when installing these patch wires. Also observe that I kept the converter chip; it's taped inside the plastic housing (you can see it as a black lump in the picture).

Figure 3.10 Patched serial adapter board.

Note that I left the DB9 on the board, even though it isn't connected to anything any more. The reason for this is that I have jammed the board back into its original housing and tightly wound it in insulation tape to keep the halves together. The DB9 acts as a strain relief and friction grip for the ribbon cable, as shown in Figure 3.11.

Figure 3.11 Reassembled serial adaptor.

There is one further thing you need to do: As previously described, the interface will let you see kernel messages, but it will not let you type anything in. There is a series resistor on the receive line, which Buffalo left off the PCB for some reason. Buffalo used a 10K resistor in the other serial line (R75), so that's what I used for R76, and it works very well. The package size is 0603; you should be able to salvage one off a junked appliance or PC peripheral if you don't keep reels of SMD parts in your lab.

Figure 3.12 is a photo of the J1 area on the component side of the Kuro PCB, showing where you need to install the resistor; it's space R76 in the photo.

Figure 3.12 Unpopulated resistor.

Now you're up and running, connect your terminal program to the USB adapter at 57600bps, 8 data bits, no parity, 1 stop bit, with no flow control. You should see all the kernel messages while the unit is booting or shutting down (and also on events such as USB hotplug). There is a getty process running on that port, so you can log in locally. The port can also be used for general-purpose communications; it is accessible as /dev/ttyS0.

With that digression out of the way, on to the design of our actual circuit. I have chosen to use an Atmel ATmega32 microcontroller as the core of the VCM. This happens to be an ideal choice for me, because I'm very comfortable with programming this micro, it's 5V-compatible (and hence easily interfaced to all the power electronics; it can be a bit irksome using 3.3V micros in projects like this), and it operates over a wide temperature range. It's also supported very well by free or low-cost hardware and software tools, and most of those tools work within Linux. Most of the same things, except for the voltage comment, are also true for the MSP430 series—if you're more familiar with that microcontroller, then you can probably port 75% of the circuit and code across without much difficulty.

Figure 3.13 shows a schematic of the circuit we'll be working with.

The ATmega32 offers a rich set of peripherals useful for this sort of project, but inevitably once you start to use those peripherals, the availability of general-purpose I/O pins shrinks.

To realize the maximum amenity from the ATmega32's peripheral set, you therefore need to categorize your I/O requirements, first according to whether a hardware feature of the microcontroller can carry out or accelerate the desired function, and second (in the case of functions that you will be implementing entirely in software), according to the specific timing limits for implementing the function. For example, something like a radio receiver input, with tight timing requirements, would be best connected directly to a micro input pin (preferably a pin with interrupt-on-change capability).

You then need to build enough I/O expansion to get around the shortage of I/O pins on the microcontroller. You can achieve this goal in many ways, of which these are the top three:

- Use external flip-flops (for example, 74HC373) to latch expansion output data and external tristatable buffers (for example, 74HC244) to scan in expansion inputs.

- Use the ATmega32's on-chip SPI module to drive external I/O expanders.

- Use the ATmega32's on-chip I²C (TWI, see sidebar1) module to drive external I/O expanders.

Each of these methods has its own distinct advantages and disadvantages that are worth exploring.

Using external logic is a simple route. It also offers the lowest possible latencies. However, the difficulty is that it directly consumes a relatively large number of I/Os. For example, if you wanted to add 24 I/Os to your device, you could add three 74HC373s and three 74HC244s. The inputs of the 373s and the outputs of the 244s would form a common data bus, routed to eight pins on the micro. But you would also need a read/write strobe line and at least three addressing pins—bringing the total required I/O count up to 12.

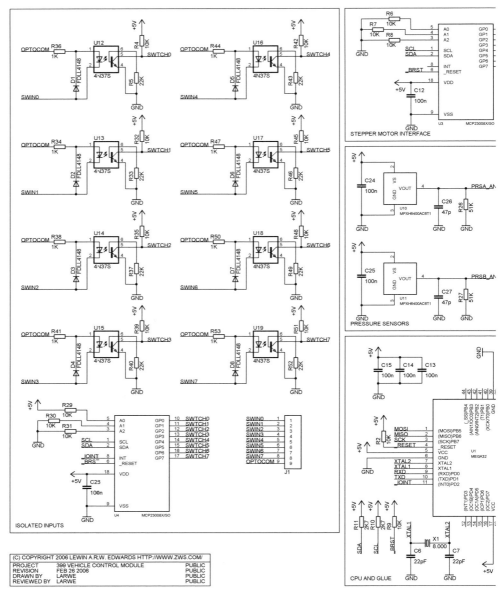

Figure 3.13 Schematic of our application.

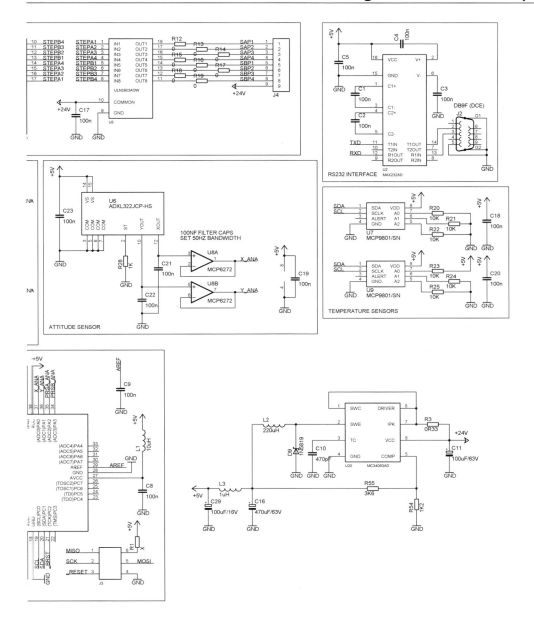

The simple external logic method also can't give you an interrupt when input states change; in some applications, that might be important. Of course, you could structure external interfaces with jellybean logic in many other ways, but it's a headache to design and debug and is often difficult to route on the printed circuit board (PCB) as well.

Serial-connected I/O expanders utilize CPU pins much more effectively, and these chips also provide interrupt-on-change capabilities on the inputs, if you should require it—they are also generally programmable, so you can remap inputs and outputs as necessary with software changes rather than needing to mess with physical wiring. All things considered, serial I/O expansion is definitely the way to go.

The two most popular low-bandwidth synchronous serial buses for intra-board communications are I²C (Inter-IC Communication) and SPI (Serial Peripheral Interface). However, some slightly confusing nomenclature is in use: Atmel calls its I²C interface TWI (Two-Wire Interface), and many other vendors refer to I²C as 2-Wire, and SPI as 3-Wire. The reason for this is intellectual property rights—I²C is trademarked by Philips, and SPI is trademarked by Motorola. Everybody who implements a compatible interface without paying licensing fees has to hide behind an alternate name. Once in a while you will encounter a truly proprietary serial bus, but it's far from common.

Now. SPI is usually described in fashionable literature as a three-wire interface, but in any system with more than one peripheral this is a bit of a smoke and mirrors act—it really requires four lines: clock, data in, data out, and select. All SPI peripherals are connected to common clock, data in, and data out lines, and each peripheral has its own dedicated select line. (The two data lines are referred to as MOSI and MISO, acronyms for Master Out, Slave In; and Master In, Slave Out, respectively.) To address an SPI device, you assert its _SS (Slave Select) pin and start clocking data out of your MOSI pin. The same clock signal pulls data out of the peripheral into your MISO pin, so after eight clocks you've simultaneously sent a byte and received a byte. The transaction ends when you deassert the _SS pin.

SPI is very easy to use (especially if you have to implement it in pure software), but the requirement for individual _SS (Slave Select) lines means that you need to add extra wiring—and an extra I/O—for every peripheral, rather than having a true "bus" architecture. It is therefore slightly harder to route complex SPI buses over a PCB.

I²C gets over this limitation and allows you to add extra peripherals to the bus simply by wiring them in parallel with existing devices. It achieves this at the expense of some implementation complexity. I²C is a two-wire interface; the signals are named SCL (Serial CLock) and SDA (Serial DAta). I²C peripherals have a 7-bit address—generally, some of the address is fixed for a given part, and a few bits are configurable by means of resistor strapping or EEPROM setup. The protocol is, unfortunately, quite involved, with numerous meaningful states and state transitions. Luckily for us, practically all of the requisite magic is implemented in hardware inside the ATmega32's TWI module; this is a case of hardware support really saving you a lot of design and debug time. All you need to do is add two pull-up resistors selected according to the load capacitance of your I²C bus and the desired maximum data rate, and wire up all your devices in parallel.

The principal disadvantage of I²C is that it's difficult to propagate over long distances through noisy environments. Because the data line (and, in multimaster systems, also the clock line) is bidirectional, you can't shape up the signals with a simple buffer. I²C was designed for, and works best in, communications on a single board or within a subassembly. It's commonly used in TV sets, PC motherboards, laptop batteries (and other smart battery types), and other consumer equipment.

I have an ulterior motive for using I²C—namely, the rich variety of peripherals available with I²C buses. Practically any sensor you can think of that might be of interest in an embedded application is available with an I²C-flavored interface—for example (and this list is far from exhaustive), temperature sensors, pressure sensors, battery charge controllers, control interfaces for image sensors, television on-screen display chips, and even complex subassemblies like flux gate compasses, gyroscopes, and GPS receivers. Although some of these sensors are available with SPI-style interfaces, I²C is more widely supported.

As you see from the schematic, I've chosen to use the Microchip MCP23008 8-bit I/O expander. This chip supports the three standard I²C data rates of 100 kHz, 400 kHz, and 1.7 MHz. Note that there are other expander ICs, such as the Philips PCA95xx series parts, that offer more bits of I/O for roughly the same price as the MCP23008. I'm using the Microchip device in this demonstration project because it's available in a dual inline package (DIP) and is therefore easier to hand-prototype; the PCA95xx parts are only available in SOIC and (T)SSOP. Moving to one of the higher pin-count devices is, however, a trivial modification, both in terms of hardware and code.

The MCP23008's outputs drive the inputs of a ULN2803A octal Darlington driver, which I use to drive two four-phase stepper motors. Excellent background reading on stepper motors can be found at *<http://www.doc.ic.ac.uk/~ih/doc/stepper/>*, along with some handy hints on how to identify steppers scavenged from computer peripherals. By the way, I have amassed a huge collection of stepper motors simply by picking up my neighbors' unwanted inkjet printers and flatbed scanners off the curb on trash day. Because it is now cheaper to buy an entire new printer with fresh ink cartridges than to buy a set of replacement cartridges, I usually see two or three junked printers per week, each one a trove of motors and mechanical parts suitable for miscellaneous experiments.

For more formal projects, you can buy new, documented steppers from numerous industrial supply sources. eBay is also a rich source of brand new steppers, usually with documentation—although you often have to buy them a tray at a time.

Note that the ATmega32 allows you to perform I²C operations asynchronously, using interrupts (though it can be quite complex to write an interrupt service routine (ISR) that covers all the bases if you're using several different I²C peripherals that have radically different protocols). The code I'm presenting here uses a simpler polled method. Note that this I²C code is not reentrant! If you plan to use the TWI module from an interrupt routine as well as main process code, you must erect suitable barriers to prevent an interrupt from attempting to use I²C in the middle of a main process I²C transaction. The firmware I've built is carefully designed to avoid this sort of contention.

Observe that the VCM talks to the SCM over a serial link. The packet format for this link is as follows.

Name	Size	Description
STRT	Byte	Start character, '!'
CSUM	Byte	Unsigned checksum of all bytes in packet from CSUM+1 to end of BODY (or, all bytes except STRT and CSUM)
TXSN	Byte	Serial number of transmission (see the following)
IRSN	Byte	'In Response To' serial number (see the following)
CMND	Byte	Command or data ID
BLEN	Byte	Number of bytes in BODY or zero if no BODY
BODY	Variable	Additional data, if required

Each end waits for a "!" character to begin buffering data. The packet is deemed complete when the requisite number of BODY bytes has been received. Bytes received beyond the capacity of the serial-receive buffer are discarded. Since the start character might occur in the middle of real data, the CSUM field is used to verify that the data block just received was in fact a complete packet.

But what are these "serial numbers" I mention? They are a means of identifying specific question-answer pairs in a transmission stream, as well as for spotting dropped packets. At power up, both the VCM and the Kuro Box initialize their internal serial number counter to 1 (0 is a reserved value). When one end of the link wants to send a packet, it sets the TXSN field to the current serial number value, then increments the serial number counter (if the counter overflows to 0, it is reset to 1). If the packet being sent requires a response, when the other end replies, it sets the IRSN field to the TXSN from the packet to which it is responding. An "originated" packet, or, one that is not a response to a request from the other side of the link, has the IRSN field set to 0.

This very simple system allows either end to keep a history of pending data requests and track which responses belong to which outstanding requests (as well as, if necessary, implementing time-outs and retries on packets that simply don't get a response).

An example illustrates this most easily. Suppose you defined command #99 to be "Get current vehicle attitude," and command #100 to be "Get current vehicle position from GPS." Furthermore, suppose that the current serial number in the VCM was 56 and the serial number in the Kuro Box was 80. Then say the Kuro Box wants to know the vehicle's current attitude and position as soon as possible. It might send the following packets in quick succession:

- A packet with CMND=100 (get GPS position), TXSN=80, IRSN=0, and no BODY

- A packet with CMND=99 (get attitude), TXSN=81, IRSN=0, and no BODY

The GPS position can take some time for the VCM to acquire, so the attitude could easily be ready for transmission before the position data. Assuming this scenario takes place, the VCM would respond with:

- A packet with CMD=99 (get attitude), TXSN=56, IRSN=81, and attitude data in the BODY. Note that the nonzero IRSN implicitly indicates that this is a response rather than a request.

- (Later, when data is available) A packet with CMD=100 (get position), TXSN=57, IRSN=80, and the position data in the BODY.

If for some reason the VCM didn't respond to one of these packets, the Kuro Box could keep a record of which serial numbers are outstanding, and after some time-out period, reissue the request. (Note that this system breaks down if the serial number wraps and comes back to the same number as the earliest pending request. For this reason, your time-out code should also trigger if TXSN wraps to the earliest pending serial number.)

Before you run the code, you should understand what it does and how to wire up your stepper motor(s). The link I mentioned previously describes in detail how a stepper motor (synchronous DC motor) is constructed and driven, but here's a thumbnail: A stepper motor has fixed coils and a permanent magnet rotor. The types of motors I designed the VCM to use have four coils (phases), each of which has one line run to a common wire and the other wire run to the outside world. (The common wire is also run to the outside world.) When a given phase is energized, it pulls the rotor toward a certain position. The four phases are physically positioned so that if you energize them in repeated sequence (1, 2, 3, 4, 1, 2, 3, 4, 1, ...) the motor will turn continuously in a given direction.

The stepper outputs on the VCM board are labeled SAP1 through 4 (Stepper A Phase 1 through 4) and SBP1 through 4 (Stepper B Phase 1 through 4). You should connect the phase lines up to the SxPx lines as appropriate, and the common line(s) from your motor(s) to the +24V line. On the VCM, I use a 9-position terminal block to take these signals off-board.

Attention! Observe the series resistors on the stepper drive lines. The zero ohm resistors I've specified are ***strictly placeholders***; you should substitute suitable current-limiting resistors based on your drive voltage and the stepper motor's coil resistance and rated current.

This code and hardware are designed to provide very simple control of two four-phase unipolar stepper motors, Stepper A and Stepper B. The step functions are driven, by default, off a 100 Hz software timer. This is a moderate speed for junkbox steppers, and a fairly safe arbitrary choice on my part. The hardware is capable of considerably higher step rates, however. You should be aware of the following two issues as you increase step rates:

- Without going into the detailed mechanics of it, a stepper can't start from zero and instantly begin operating at its maximum rated step frequency. You need to implement "acceleration profiles" if you intend to get the most out of your motors. In brief, this means that if you need to make a speed change on the motor, instead of just jumping to a new step rate, you ramp the step rate toward the target speed. Acceleration profiles need to be calibrated for the motor and the mechanical load being driven.

- As you increase step rates, torque decreases. Cunning hardware designs are necessary to squeeze the maximum possible performance out of a given motor.

This particular design is nowhere near to pushing the envelope in either of these respects—you have plenty of headroom to increase step rates, for instance.

Referring to stepper.c, you'll see that the software keeps track of the absolute position of each stepper, checking it at the step speed interval. If the position ordered by the Kuro Box is not the same as the current position, the appropriate motor is stepped either forwards or backwards. The serial interface code listens for "seek to..." packets and updates the target values appropriately.

The VCM code assumes that the power-up position of each stepper motor is its known absolute zero position. Finding the zero reference is normally accomplished by having a hardware zero position detector attached to the motor's drive shaft (or the mechanism it operates). For example, in floppy disk drives, the track 0 sensor is an opto-interrupter gate, or (in older drives) sometimes a microswitch that the head engages when it is stepped to the track 0 position.

The two packet types supported by the VCM firmware in the vcm_399 directory are CMD_STEP_A and CMD_STEP_B (defined in vcmpacket.h). If you build the Kuro Box demo program in the scmd directory, you'll see the Kuro Box printing out status info from the VCM and, once every four status packets, sending the VCM commands to seek both stepper motors to random locations. Please note that the Kuro Box side of the code is quick and dirty; it's a proof of concept applet, as we're focused on the VCM side of the design here.

Here's some sample output from the scmd program:

```
root@KURO-BOX:/mnt/share/article8/scmd# ./scmd
IBM developerWorks Kuro Box to VCM Demo Applet #2—Stepper Demo
Waiting for VCM to start sending...
Rx packet (TXSN 0x01, BLEN 0x05)—MTIME 0x000005b8, FLAGS1 0x00
Rx packet (TXSN 0x02, BLEN 0x05)—MTIME 0x00000b70, FLAGS1 0x00
Rx packet (TXSN 0x03, BLEN 0x05)—MTIME 0x00001128, FLAGS1 0x00
Rx packet (TXSN 0x04, BLEN 0x05)—MTIME 0x000016e0, FLAGS1 0x00
Tx 2 packets: CMD_STEP_A->00000167, CMD_STEP_B->FFFFFF97
Rx packet (TXSN 0x05, BLEN 0x05)—MTIME 0x00001c98, FLAGS1 0x00
Rx packet (TXSN 0x06, BLEN 0x05)—MTIME 0x00002250, FLAGS1 0x00
Rx packet (TXSN 0x07, BLEN 0x05)—MTIME 0x00002808, FLAGS1 0x00
Rx packet (TXSN 0x08, BLEN 0x05)—MTIME 0x00002dc0, FLAGS1 0x00
Tx 2 packets: CMD_STEP_A->FFFFFFAF, CMD_STEP_B->0000004A
Rx packet (TXSN 0x09, BLEN 0x05)—MTIME 0x00003378, FLAGS1 0x01
Rx packet (TXSN 0x0a, BLEN 0x05)—MTIME 0x00003930, FLAGS1 0x00
Rx packet (TXSN 0x0b, BLEN 0x05)—MTIME 0x00003ee8, FLAGS1 0x00
Rx packet (TXSN 0x0c, BLEN 0x05)—MTIME 0x000044a0, FLAGS1 0x00
Tx 2 packets: CMD_STEP_A->FFFFFED7, CMD_STEP_B->FFFFFF46
Rx packet (TXSN 0x0d, BLEN 0x05)—MTIME 0x00004a58, FLAGS1 0x00
Rx packet (TXSN 0x0e, BLEN 0x05)—MTIME 0x00005010, FLAGS1 0x00
Rx packet (TXSN 0x0f, BLEN 0x05)—MTIME 0x000055c8, FLAGS1 0x00
Rx packet (TXSN 0x10, BLEN 0x05)—MTIME 0x00005b80, FLAGS1 0x00
Tx 2 packets: CMD_STEP_A->FFFFFE0E, CMD_STEP_B->000001E3
Rx packet (TXSN 0x11, BLEN 0x05)—MTIME 0x00006138, FLAGS1 0x02
Rx packet (TXSN 0x12, BLEN 0x05)—MTIME 0x000066f0, FLAGS1 0x00
Rx packet (TXSN 0x13, BLEN 0x05)—MTIME 0x00006ca8, FLAGS1 0x00
```

Observe that the VCM informs the Kuro Box if it's busy stepping every time the VCM sends status; this information is contained in FLAGS1 in the CMD_STATUS_REPORT packet. FLAGS1 bit 0 set means that Stepper A was running when the status packet was generated; likewise, bit 1 means that Stepper B was running.

So now you have most of the makings for a set of dive planes and a rudder. But in order for this device to be very useful, it also needs sensors.

We will start with the humblest of sensors: a simple switch. You might need a couple of these to delimit the travel of your rudder, or if you're making something other than a robot, maybe you want some front-panel buttons (though if that's all you want, you can attach these to the PowerPC® in better ways). Switches are so simple that in fact the only reason I'm bringing them into the discussion here is so I can gently lead into a slightly more complex I²C configuration than the circuit you met last time. I'm going to assume that these switches will need to be read relatively infrequently (say, in the neighborhood of 50–60 Hz). In keeping with the I/O usage philosophy I espoused earlier, this slow rate can safely be marooned on the other side of an I²C I/O expander. However, we used up an entire expander on the stepper motors. How can you add another block of I/O?

Fortunately, all you need to do is wire another MCP23008 onto the I²C bus. The chip has a seven-bit I²C address; four bits of this address are fixed, and the remaining three can be configured in your external circuit by means of external strapping resistors on the A0, A1, and A2 lines. This allows you to connect up to eight MCP23008s to a single I²C bus without any ugly complexities. (This is a very common sort of arrangement on I²C peripherals, by the way—for cost reasons, manufacturers will very rarely bring out more than a few address pins.) The expander that drives the stepper motors is at address 0 (its full binary address is 01000000, where the bottom bit is actually the read/write flag). We'll add a second expander at address 1 (again, the full binary address byte for this second chip is 01000010). The code in i2c.c handles the translation from logical address (0–7) to physical address byte for you.

In the interest of full disclosure, if you look at the way I scan these switches, you'll see a rather large cheat in my debouncing algorithm. I simply scan at periodic intervals, and check for a change in input state. If there is a change, the new data remains pending until the next scan interval. At that time, if the change in state is still the same, the new data is latched into a status buffer. This is not a very advanced debouncing method by any stretch of the imagination, but it performs adequately in lab conditions (at least, with reasonably well-behaved switches). If you're interested, look at *<http://www.ganssle.com/debouncing.pdf>*, where you can read a very detailed article on debouncing techniques, accompanied by reams of actual real-world data. If you want to implement a more advanced debouncing method, I've wired the interrupt request line from the MCP23008 into one of the GPIOs on the microcontroller. Since this line can be configured as an open-drain output, you can add more I/O expanders and simply connect all their interrupt lines to the same point for a wired-AND configuration. An external pullup is not necessary, as the ATmega32 has on-chip pullups.

One further subtlety I've added to this circuit is optical isolation of the input lines, accomplished very easily with a standard six-pin optocoupler and a couple of resistors. This additional circuitry serves two purposes: first, it protects the microcontroller from outrageous external events such as miswired connectors or electrostatic discharge, and it also provides a kind of level-shifting capability; you can interface practically anything to the input side of the optoisolators. You'll be very grateful of this isolation circuitry if you ever accidentally tap an unregulated battery line onto one of the inputs; it's much easier and cheaper to desolder an optocoupler than to replace the micro! Please note that if you want to get the full amenity of the ESD isolation, you will need to ensure that whatever you have outside the VCM box has a separate (or at least, isolated) power supply from the VCM. Otherwise an ESD event on your external hardware will propagate into the VCM through the common power rail, largely negating the benefit of the optocouplers. Note also that the series resistor I selected for the optocoupler LEDs is correct for a +12V external supply; you will need to tweak this if you run the common anode line to a different voltage.

Next, you probably want to monitor some temperature points in the vehicle. In the real E-2 submarine, I'm interested in several temperatures—the two drive motors have a temperature sensor apiece, as does each battery. Another temperature sensor is thermally connected to the external environment so I can have an idea of what the water temperature is like. For the purposes of this article, I'll only implement two of these sensors. I'm using the Microchip MPC9801 12-bit I^2C temperature sensor for this application; adding more measuring points is simply a matter of mounting the sensors where you need them, and wiring them onto the I²C bus. As with the MPC23008, there are three bits of user-configurable address; the overall device address byte is 1001xxxR, where xxx are address pins A2 through A0, and R is the R/W bit. Observe that there is no possibility of address collision with the MCP23008s, no matter what A2/A0 combinations you select for either. As a matter of interest, if you refer to the datasheet for the MCP9800 series, you'll see that the part is also offered in a couple of flavors that are packaged in a five-pin SOT23 (surface mount transistor) form factor. This tiny package doesn't have space for external address select pins, so Microchip offers one flavor that's hardwired at address 000 and another at address 101 (binary). Again, this practice is quite common with I²C sensors.

The MCP9801 also features a thermostat mode (this option is frequently provided on digital temperature sensor devices). This feature consists of a single open-drain output that goes active if the sensed temperature falls outside programmable limits, and it operates completely independently of the I²C logic, so you can use it as a hardware fail-safe even if the microcontroller crashes. Perhaps more important, you can even shut the micro down completely to save power, and leave the temperature sensor IC running to shut things down and wake the micro back up if unreasonable temperature excursions occur.

In the real E-2, I use the MCP9801's thermostat feature to kill charge current if the batteries get too warm, to stop the drive and compressor motors if any of them exceeds a nominal temperature threshold, to pause the compressor while filling the high-pressure air bottle, and to turn off some high-intensity halogen lamps if the temperature of the hull area surrounding the lamp's reflector rises too high (this might indicate heavily sedimented water, but the situation I'm most concerned with is turning on a lamp facing straight down into mud—those lamps generate an incredible amount of heat). Note, however, that although this feature is usually described as a "thermostat" in other vendors' datasheets, and I use the word freely in this text, this part is not really suited to drive a thermal load without external intelligent assistance. The temperature alert feature, as Microchip terms it, is designed to provide a cutoff or warning signal, rather than a completely unsupervised process control input.

Next, you need to be able to sense a couple of pressures. The part I've selected for this is a Freescale MPXH6400AC6T1, which can measure from three to

58 PSIA and has an integrated hose barb. It is intended for automotive applications, but works well in the moderate pressure ranges encountered in the E-2 project. Again, in the real submarine quite a few spots need to be sampled; I'm interested in ambient pressure as a way to gauge the vessel's depth, as well as various pressures in the air lines leading from the high-pressure air bottle to the ballast tanks, as well as a couple of pneumatic linear actuators. The preceding circuit only implements two sensors, but again this can easily be extended to any number your application might require.

Note that the MPXH6400A series is only characterized for dry air use. You can, however, use it to measure external water pressure by using an air bubble behind a flexible diaphragm. That phrasing sounds really scientific and technical, so I'll freely admit that the "flexible diaphragm" in question is actually a plastic soda bottle. I drilled a hole in the bottom and epoxied a tube into the hole. I also glued the cap on tightly with more epoxy, after first removing the rubber seal. This arrangement has been tested at pressures up to two atmospheres, and it would probably withstand quite a bit more.

These pressure sensors provide analog outputs. We read them at a fairly low sample rate using the analog-to-digital channels of the ATMega32L. Some basic software filtering removes noise; we don't expect the values here to change very rapidly. You can find the relevant code in main.c.

The last and most complex sensor we'll use is an Analog Devices ADXL322 MEMS (Micro Electromechanical Systems) accelerometer. MEMS is an exciting technology that straddles the border between "really, really small machines" and nanotechnology. The most common MEMS devices you'll encounter in robotics work are accelerometers and gyroscopes; various vendors including Analog Devices, Freescale, ST, and Kionix® (among others) offer these sorts of parts. (By the way, the pressure sensor we're using is also a MEMS device.) If you browse the Resources in depth, you'll see quite a few other very interesting MEMS devices on the market, both sensors and actuators. I'm particularly intrigued by the possibility of building an electric motor the diameter of a human hair, though I can't yet think of a use for this device in any project I'm working on.

The ADXL322 is a two-axis ±2G accelerometer. This means that it can measure acceleration in two dimensions, and these two dimensions are at right angles to each other. The sensor output saturates at ±2G. Other typical ratings for accelerometers are ±5G and ±10G. Parts with higher acceleration ratings are used in applications such as car airbags, which need to be able to discriminate between high-speed and low-speed collisions. (Second-generation and later car airbags have different firing behavior depending on the severity and direction of the collision.)

The 2G accelerometer I'm discussing here is typically used for measuring roll and pitch of a vehicle or perhaps a video game controller. For example, you might use it in an auto-leveling circuit for a model aircraft. The accelerometer would be mounted with one axis—the X axis, without loss of generality—parallel to the stern-to-bow line of the craft, and the other axis (Y, in this case) running from port to starboard. (See Figure 3.14.)

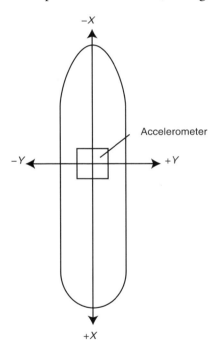

Figure 3.14 Accelerometer mounting.

The ADXL322 provides you with two analog voltage outputs corresponding to the X and Y acceleration vectors. When the device is parallel to the Earth's surface, both analog outputs are mid-scale. As you tilt the device toward the $X+$ direction, the X output gets closer to rail voltage; tilt it back toward the $X-$ direction and the output heads toward 0V; similarly for the Y axis. In general, the roll and pitch (when the sensor is mounted as previously described) are given by the relations:

$$pitch = \sin^{-1}(X)$$

$$roll = \sin^{-1}(Y)$$

where X and Y are numbers scaled from the voltage ranges provided by the chip to a −1.0 to +1.0 scale. Due to motion or noise, X or Y can exceed 1.0; you need to allow for this possibility.

An important point that might not immediately be obvious: A two-axis accelerometer can only tell you your vehicle's acute angle off vertical in the X and Y directions. It cannot completely resolve this information into a unique vehicle orientation. Consider the two situations in Figure 3.15, where I roll a vehicle through 15 degrees and then through a further 150 degrees (note that these diagrams represent a stern view of the vehicle).

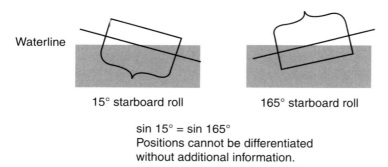

Figure 3.15 2D accelerometer limitations.

As you can see, the roll axis output is the same for both situations. The z-axis, if I had a way to measure it, would have changed sign, but the two-axis device simply cannot differentiate between the two possibilities.

Another point that sometimes seems very difficult to communicate is the following: An accelerometer only measures a single acceleration vector (in this case, I have the vector decomposed into two components; a three-axis accelerometer would give a third component, but the net result is still a single vector). Gravity is one component of this acceleration. Your hand pushing the device across the table might be another. A rocket motor launching you, the table, and the device into space would just be another component of the acceleration. You ***cannot*** separate out these components just by looking at the output of the accelerometer. In other words, if you have the accelerometer mounted as I previously described, the best you can do—even with a three-axis accelerometer—is to obtain a single vector that summarizes all linear acceleration forces acting on the vehicle.

Due to this fact, and also the lack of a *z*-dimension and the low-maximum-G sensor being used, the circuit I'm discussing here is totally unsuitable for inertial navigation. Building so-called dead reckoning inertial guidance circuits and developing the software for them is very challenging. To establish the motion vector of an object from a historical record of its acceleration vector, you need to sample rapidly and (numerically) integrate over time. To calculate the net displacement (position) of the object, you need to integrate the motion vector again. Errors in these processes are cumulative.

The consumer application where you are most likely to encounter such circuits is in very high-end GPS receivers. These devices try to keep track of their position using the GPS signals as far as possible. When the satellites are temporarily occluded (for example, when you drive into a tunnel), the device keeps your position up to date, with reduced accuracy, using inertial navigation.

The MEMS device is read through the Analog/Digital Converter (ADC) channels, just like the pressure sensor. A slight nuance here is that due to impedance matching issues, you need to use an op-amp configured as a simple voltage follower between the sensor and the micro. You'll observe a few calibration constants in the code; you need to calibrate the zero reference (flat on the table) and the ±1.0G voltages for both outputs. These constants will vary from unit to unit, due to differences in wafer orientation of the MEMS sensor inside its packaging, different mounting angles of the package on the PCB, and other random factors. The usual way of calibrating these devices is with a reference platform. (A piece of plywood with adjustable-height feet—or, large bolts—in each corner and a spirit bubble on it, is an adequate reference platform.) You calibrate the zero position with the vehicle on the table; you then roll the vehicle through ±90 degrees and pitch it through ±90 degrees to calibrate the range limits. You can store the constants thus measured in EEPROM; they will be valid unless the device is serviced and the orientation of the accelerometer with respect to the vehicle exterior is altered.

We've talked a lot about the hardware; what about the firmware? To flash new code into the AVR, you will need an in-system programmer such as the Atmel AVR-ISP cable, the STK500 development board, or a third-party tool. You'll also, obviously, need a compiler and linker, for which I chose to use avrgcc. Although I work in Linux, you can equally easily build the AVR GNU tools in Windows using the cygwin emulation environment, available freely from *<http://sourceware.org/cygwin/>*. Cygwin runs under Windows 95/98/Me/NT/2000 and XP (and presumably Vista, though I haven't tried this). It emulates most of a Linux system and makes it possible to run a large amount of UNIX® software (including X11 software) directly within Windows. If you wish, you can also use the precompiled WinAVR tools; I generally prefer to build my own toolchain, but if you're looking to get started quickly, WinAVR is the speedy route.

If you're rolling your own, your first step is to download `gcc` and `binutils` from *ftp.gnu.org*, and `avr-libc` from *<http://www.nongnu.org/avr-libc/>*. First, unpack build and install `binutils` (assuming you're using version 2.16, which is the version I am currently using for AVR development):

```
tar zxvf binutils-2.16.tar.gz
cd binutils-2.16
./configure --target=avr --program-prefix="avr-"
make
make install
```

Next, unpack, configure and install `gcc` (again, assuming you're using 4.0.2):

```
tar zxvf gcc-4.0.2.tar.gz
mkdir gcc-build
cd gcc-build
../gcc-4.0.2/configure  --target=avr  --program-prefix="avr-"  --
enable-languages=c
make
make install
```

Finally, unpack, configure and install the `avr-libc` runtime library (I'm using version 1.4.4):

```
tar z6xvf avr-libc-1.4.4.tar.gz
./configure --build='./config.guess' --host=avr
make
make install
```

(*Note*: The ' character is the single quote, not the apostrophe. This character is found under the tilde (~) on a standard U.S. 104-key keyboard.)

By the way, if you downloaded the .bz2 versions of these files, then instead of using the syntax:

```
tar zxvf filename.tar.gz
```

you should use:

```
bunzip2 -c filename.bz2 | tar xvf -
```

What you just went through is merely a quick HOWTO-style thumbnail of how to build and install the GNU toolchain components; for more information, please refer to the online documentation or simply obtain a copy of my first book. (I promise you that I'm not trying to shill that book. I just get very tired of rewriting the material that I put in that volume—it's much easier for me just to write it once and point people to where I put the information!)

You might also want to install a command-line programming utility. However, take warning! From time to time, Atmel changes the communication protocol used by its serial programming tools. When you download and run a new version of AVR Studio for Windows, it will often prompt you to update your STK500's firmware without warning you of the consequences. One of these "flag days" occurred very recently, and it has completely broken third-party support for the STK500. The only open-source program I'm aware of that's currently capable of using this new STK500 protocol is the latest beta of avrdude; older programs such as uisp are effectively useless. If you have a working system based on third-party tools, I advise that you never run AVR Studio when your STK500 (or AVRISP cable) is connected!

To build the firmware for the AVR, assuming you already have avrgcc properly installed, simply extract the source tarball (available at *<http://www.zws.com/publications/downloads/ibm-article9.tar.gz>*—it's not an efficient use of space for me to print it here) and run make in the vcm_399 directory. Take a moment to examine the structure of the program. In particular, note how the interrupt syntax works. The avrlibc library by default vectors all the AVR interrupts to a do-nothing handler. To revector an interrupt, rather than writing a normal function name, you use the SIGNAL macro with an argument naming the interrupt to be vectored—followed immediately by the code block to be inserted. For instance, in serial.c you'll see the serial receive interrupt declared as:

```
SIGNAL(SIG_USART_RECV)
{
        // code goes here
}
```

WARNING: The signal name definitions are not 100% identical between different AVR variants, and sometimes their meanings can be a little confusing. Therefore, C code that uses interrupts is not guaranteed to be portable, even if both AVRs in question share the interrupt-generating hardware feature you're interested in. If your interrupts are inexplicably not firing, look at the appropriate header file in /usr/local/avr/include/avr (assuming you used the default install location for everything) and verify that you used the correct interrupt name for the particular chip you're using.

Because of this potential for confusion, when you're bringing up a new design, I strongly advise using a spare I/O as an interrupt flag. Before you start puzzling out obscure interrupt priority questions, just do a quick sanity check—force your interrupts to fire, and verify on the scope that you're reaching the correct code point.

Another interesting feature you might want to look at is the EEPROM handling code in eeconf.c. Although nothing is stored in there for this little demo applet, in the final product the EEPROM is used for some important calibration constants (accelerometer zero roll/pitch values, for instance). The code in eeconf.c implements a simple checksummed redundant configuration scheme. The avrlibc library provides a convenient framework for polled EEPROM read/write code. If you have higher performance requirements and want to use interrupt-driven writes, you have to roll your own.

A handy hint: To get a quick look at ROM and RAM utilization in your avrgcc program, use `avr-objdump` to look at the section headers—in this case, `avr-objdump -h 399.elf` will show you what you need to know (look at the Size column, and ignore .stab and .stabstr, which are symbol tables that don't get uploaded to the chip).

Speaking of memory, you should be aware that avrlibc isn't very efficient size-wise. For instance, if you use printf, you'll pull in a huge amount of code (this is a common problem in embedded systems). There are various stripped-down `printf` functions you can use, or you can simply implement by hand the bare functionality you need. Since my application doesn't need to transmogrify much output into human-readable formats, I chose the latter route; look at utils.c for examples of that code. (These functions are not used by the code; I just include them by way of completeness.)

If you've followed along in this section (and hopefully looked at the sample sourcecode and Makefile), you'll have a good idea of the kind of thought and design process that goes into a fairly well-specified, but informal project. Adding work like this to your portfolio will greatly enhance your employability.

4

Teaching Yourself, Top-Down (Large Embedded Systems)

4.1   Target Audience

The previous chapter dealt with people who have hardware experience and want to start learning about microcontroller programming, or a "bottom-up" approach to embedded engineering. At the other end of the spectrum, we find people who have considerable experience programming application software in high-level languages, and who now want to extend their reach into embedded systems. These people will typically have what I would describe as "IT" qualifications (database programming, HTML design, Java development and so forth), rather than engineering experience. Computer science majors quite frequently fall into this category.

There is an immediate difficulty in adapting these people to embedded environments, and this is that pure software projects for mass-market operating systems (Windows, Mac OS, and so forth) are typically designed with an acceptance of enormous variations in performance margins due to variations in customer hardware. Furthermore, PC hardware is sufficiently expandable and inexpensive that it is realistic for an application software developer to require that the user provide special hardware features such as additional memory, 3D accelerated graphics, and so on. Neither of these two assumptions are remotely acceptable in embedded environments.

The embedded software developer *must* be able to:

a) characterize and control the resource utilization of the software exactly. This includes being able to specify how much RAM and secondary storage space the software will require under all conceivable execution conditions, with explicit safeguards in place to prevent the software from overrunning those limits under unusual input circumstances. In most cases, it is also necessary to provide some guarantee of real-time performance (though the limits here are often loose enough that people don't need to take explicit notice of them).

b) develop the software in such a way that it utilizes available system resources efficiently, and

c) design the software to perform deterministically under specified input conditions.

None of these skills are apparently in great demand in modern consumer applications software development;[1] hence, they are not well taught in computer science degree programs. Also observe that the scope of the word "software" in this context explicitly includes the operating system running on the target device; by contrast it isn't normal for application developers on PCs to have to guarantee the behavior of the underlying operating system.

Ten or twelve years ago (as I complained in the introduction to this book), this gulf between embedded development and mass-market software development was considerably smaller; programmers who cut their teeth writing games and other software for the Commodore Amiga or (even more so) its 8-bit antecedents such as the Sinclair ZX Spectrum and Commodore 64 were well-prepared for life developing real-time embedded software.

This chapter will introduce you to several of the options for developing high-end, 32-bit (or even 64-bit) embedded systems. Pay attention to the "high-end" qualifier there. Some 32-bit microcontrollers (ARM and SuperH, for example) are available in exceedingly cut-down variants, usually with no external memory buses, intended for low-cost single-chip systems. Other 32-bit cores, many of them proprietary, are built into ASSPs such as those used in DVD players. Both these categories of parts and applications more properly fall into the province of Chapter 3. In this chapter I'm very specifically focusing on complex embedded systems where the application layer is considerably abstracted from the hardware.

---

[1] Engineers generally have a very special view of the deplorable state of consumer software development. This is one reason why you'll find a disproportionately high number of engineers running open-source operating systems. They've got bugs, just like commercial software, but at least there's the *opportunity* for a user to fix the bugs (even though most engineers would never have enough free time to start tinkering with that code).

These applications would typically involve complex user interface code, often with a GUI. Examples of such systems might include an ATM, an electronic storefront sign driven by an embedded PC, a PDA, an automated movie ticket vending machine, or a computer-driven airport/railway station destination board. Note that I am specifically *excluding* hard real-time systems that simply happen to have such high performance requirements that they need a 32-bit microprocessor.

In a similar fashion to the previous chapter, I'm going to give you a brief taste of the capabilities of a few popular high-end platforms. Again, this is not supposed to be an introduction to actually developing for these systems. The goal is to elucidate for you some of the strengths and weaknesses of these more powerful microcontroller families. This will help you to make an informed choice as to where you can begin experiments of your own.

4.2 Embedded x86 Solutions

Developers who are most comfortable with application development under Windows or Linux will quite likely gravitate toward Intel-compatible x86 systems out of sheer familiarity. While x86 has some distinct disadvantages in many embedded applications, it is not to be despised when used in environments that benefit from compatibility with off-the-shelf PC hardware and software.

Let me digress here for a moment to explain why I selected the previous phrasing; it wasn't arbitrary. Simply put, x86 is not a very good candidate for many embedded systems. The x86-compatible extended family is (with a few exceptions that you can safely ignore) only employed as part of a more or less wholly PC-compatible hardware architecture. These processors and their support chips are large and *extremely* energy-hungry; active cooling is almost universally required in x86 designs, and mains power is preferable. The dominant engineering factor steering x86-based designs is the baggage required by backwards compatibility (both in the CPU core and other support circuitry on the motherboard), and most of the people reading this text are doubtless aware of the history behind this statement. Vast efforts have been invested—some uncharitable folk might say "squandered"—to modernize the CPU and overall system architecture, but even the current 64-bit architectures are constrained to some degree by legacy considerations. There are relatively few true system-on-chip offerings based around an x86-compatible core, and thus significant external circuitry is always required in x86 systems.[2] It should also be noted that, unlike the vast majority of embedded devices, x86 chips lack on-chip JTAG or other hardware debugging interface support. This makes debugging critical low-level functionality relatively difficult.

x86 shines, for the embedded developer, in a few situations:

1. You need to hit an extremely aggressive price-performance point for a very complex high-performance hardware design in low production volumes. Because most of the parts in an embedded x86 system are off-the-shelf consumer PC parts, the unit pricing is lower than it would be for parts specifically designed for niche markets. Your 100-piece order of single-board computers enjoys lower overall component pricing by riding the massive volumes purchased by Dell, Gateway, and so forth.

2. You're building your product around a bunch of functionality that is available ready to run in an off-the-shelf operating system like Windows XP, and your end product doesn't need to have extremely tight real-time characteristics.

3. You need to have a very quick design cycle time—you want to develop a mockup of your code on a regular PC, then transfer it across to the target system with the minimal possible amount of porting and debugging time. This positive factor is enhanced considerably if you have access to a stable of competent application-level programmers.

Embedded x86 systems take a variety of forms. The most easily recognizable method of embedding an x86 processor is simply to take an off-the-shelf PC and put it inside a cabinet containing whatever it is you want to control. The self-service photo scanning and printing stations often found in pharmacies and department stores are built this way, and they are perfectly legitimate embedded systems. One of the principal advantages of this approach is that you can keep your application—both from the software perspective and the mechanical assembly perspective—more or less entirely isolated from the specific piece of

---

2   There are exceptions to this rule. The i386EX and some 80186 variants, for instance, are not terribly difficult to design around. For the purpose of this chapter, however, I'm discussing strictly higher-end x86 systems. At the performance level occupied by those low-end processors, there are much better non-x86 alternatives.

hardware in your box. This is a Good Thing, because consumer PC components tend to have very short production lifecycles compared to parts specifically intended for embedded applications.

At the opposite end of the scale, it is of course possible to build your own entirely custom board around an x86-compatible processor. The difficulty level of such a task is depressingly formidable, however—it's a job really best left to companies that specialize in this sort of design work. Even just to take a semiconductor vendor's cookie-cutter reference schematic and lay it out on a PCB shaped to fit your own space requirements is assuredly not a simple task. At the very least, you need to license a BIOS and tweak it to initialize the memory controller correctly for your board. There are subtle timing parameters that must be adjusted to account for layout-specific issues in high-speed memory subsystems—essentially, you need to skew the timing on each data line to account for the fact that the trace lengths and delays are different. Testing this and finding a set of timing values that works with all the memory configurations you intend to support is quite challenging; it's difficult to get something that works at all, and *extremely* difficult to certify that your design is going to work correctly across all combinations of temperature, installed memory type, ambient RF noise, and so forth.

In between these extremes, there are a variety of off-the-shelf x86-compatible single-board computers, backplane-based systems and processor/logic core modules designed for integration with custom mainboards.

The most popular mainstream x86-compatible processors come from Intel, AMD and Via Technologies. (National Semiconductor used to build a range of Pentium-like processors collectively called Geode, but this product line was sold to AMD. You'll still see chips with the National Semiconductor logo on them.) The other main contender is Transmeta™, which makes some innovative low-power-requirement x86 emulator chips (Crusoe™ and Efficeon™) used in some consumer laptops and single-board computers. AMD's and Intel's range of consumer chips (of the type you'd find in normal desktop and portable PCs) are the best performers; as a brutal generalization, the more oddball parts such as Geode, and the CPUs from Via™ and Transmeta, are designed with specific goals in mind; power efficiency (Via Eden™, Transmeta), low embedded system cost for set-top box and Internet Appliance (IA) applications (this is where Geode was originally intended to find its niche), and so on. ***Be prepared to test your*** *application on several hardware mixes before committing to a hardware vendor*. It can be close to impossible to gauge how well a particular requirement will be met on a specific PC-compatible platform; consumer benchmarks are naturally skewed toward consumer activities such as office applications and games, which probably don't accurately reflect most embedded application needs.

The cheapest and in some ways most flexible method of embedding an x86, assuming that you're not just going to dump a complete PC into your appliance, is to use a PC motherboard with a standard form factor as the system core. Modern motherboards, in case you haven't disassembled a PC recently, bring all their external connectors for onboard peripherals (video, serial, USB, audio, and so forth) to the back of the housing. The motherboard ships with a layout-specific connector plate that fits into a standardized hole on the back of the case, and seals the connector area against RF leakage. Since the mounting screw positions and the size and shape of this connector plate area are standardized,[3] you can easily develop a housing that will accept almost any modern motherboard.

The disadvantages of this method are, however, show-stoppers for many applications. Possibly the worst downside is that in order to drive a PC motherboard you need a relatively complicated and expensive power supply. The smallest readily available power supplies that have all the standard PC motherboard connections are intended for 1U rack-mount server cases (approximately 16.8" × 25.6" × 1.7" W × D × H); the power supplies are typically in the region of 5.5" × 10" × 1.5" with some protrusions. This immediately sets a lower boundary on the overall size and shape of your product; couple in the relatively large size of a standard PC motherboard and your housing design is going to wind up not too radically different from a normal slimline PC case. You also have to deal with a host of other issues: Cooling can be a problem in these confined spaces; fans on the CPU and in the power supply are noisy and constitute potential failure points; and as supplies of a particular motherboard fluctuate, you'll need to keep altering your software bundle so your driver set matches all the different hardware you're shipping. This can also make end-user upgrades problematic, since it can be difficult to keep your upgrader totally generic while you're making all these hardware changes under the

---

[3]   Visit and bookmark *<http://www.formfactors.org/>* for all the details. You'll need it if you're building PC-based appliances; they have specifications on everything from power supply requirements to screw hole sizes.

hood. It's really irksome to have to ask users what flavor of hardware they have and keep separate upgrades available for those different versions.

The next most flexible option is to pick a specific motherboard that is still fairly generic, but is intended for or at least widely used in embedded applications. Via's motherboard arm *<http://www.viavpsd.com/>* is probably the industry leader here; they led the wave by pioneering the Mini-ITX form factor with their Epia series of motherboards, and they now have several different form factors specifically designed for embedding. (The Epia range virtually invented the category of homebrewed PC-based media players and similar embedded applications.) The downside to choosing these boards is that you're definitely locking yourself into one vendor, albeit one with a fairly strong commitment to keeping their old boards in production for a reasonable lifespan. The advantage is that these boards are smaller than most normal PC motherboards, so you can make your housing smaller; you can also calculate the power requirements more finely and potentially use a smaller, lower-current custom-designed power supply, since you don't have to worry about the possibility of switching in a much greedier motherboard at the next production run. Some boards based around the Via Eden CPU don't even require active cooling on the processor.

The big advantage of going with the "consumer embedded motherboard" route is that you don't need to worry about your motherboard being discontinued; you can buy whatever board is cheapest at the time you do your production run (physical dimensions and cooling requirements permitting). You've got insurance against the nightmare of trying to source an obsolete board while you frantically redesign your housing to match the next best choice.

The next step beyond the consumer grade is to use a single-board computer. These are available in a wide variety of form factors. The vendors with which I have the most experience are Advantech *<http://www.advantech.com/>*, BCM Advanced Research *<http://www.bcmcom.com/>* and ICP America, Inc. *<http://www.icpamerica.com/>*; there are numerous others, of course. Note that many of these vendors are simply labeling and reselling products from other people.

Common form factors you'll encounter in this market space include:

- CPU cards (ISA or PCI, usually) intended for connection to a backplane. Note that a frequently asked question is "Can I build a multiprocessor system by putting several CPUs onto one backplane?" The answer to this question is "probably not"—many of these cards are designed to be the sole master of the backplane; the other slots have to be empty, or occupied by standard PC peripherals such as network cards and so forth. Sometimes the backplane is only for power distribution, in which case you can put in several processors—but you need to network them somehow (Ethernet, usually) and develop your own multiprocessing communications software if you want to use the whole thing as a number-crunching engine. A very few CPU cards are designed to fit into a multimaster environment and talk to each other over the backplane; again, however, special software is required to get it all to work. These cards aren't usually intended for distributed processing applications, though; they're typically meant for telecommunications and network routing, where you have a bunch of low-bandwidth interfaces leading into each individual CPU card, and bulk traffic being routed between interfaces using the fast backplane.

- 2.5" biscuit PCs. These are designed to have the same footprint as a 2.5" hard disk. Due to the extremely confined space, these boards usually require custom cable harnesses for all of the connectors. They are also usually built around fairly cool-running (i.e., slow) processors to avoid thermal problems.

- 3.5" biscuit PCs, designed to match the footprint of a 3.5" hard drive or floppy drive. The sweet spot for price versus performance appears to lie in this form factor. Boards are available from the i486-class through midrange Pentium® class. These boards will typically have one or two Ethernet MACs, VGA output, one or two IDE channels, DiskOnChip® or CompactFlash® boot capability, and various other interfaces. The 3.5" form factor is not really large enough for a normal ISA or PCI expansion bus, so most 3.5" SBCs have a PC/104 interface (this is essentially ISA, brought out to a pattern of 0.100" headers in such a way that you can stack peripheral boards on top of the CPU to achieve various combinations of functionality). RAM in these boards is almost always provided by way of a laptop-style SODIMM socket.

- 5.25" biscuit PCs. This form factor appears to be in decline, though products are certainly still available. The functionality you'll find in these boards is similar to what you'll find in 3.5" SBCs, except that due to the larger board area you'll typically find a PCI slot and perhaps some integrated RAM onboard.

There are several advantages to using, say, a "biscuit" PC. The first of these is the power requirements. Most of these boards run off a single +5V rail, though some require +12V as well. You can provide this with a relatively inexpensive single- or dual-voltage open-frame power supply, or you can design your own supply if necessary. This also simplifies the job of maintaining battery backed-up operation, if your system design requires it. Second, the form factor is obviously much more amenable than a standard PC motherboard to integration inside a nice small box. Third, vendors of industrial PCs typically provide things you won't find in the consumer PC market, including: guaranteed operational temperature ranges, explicit support for Linux, longer product lifecycles with advance end-of-life notifications (lifespans of these boards are often four to eight years compared with as little as six months for consumer products), onboard watchdog timers, RS-422/RS-485 interfaces as well as regular RS-232, and more.

If you choose to build your application around an embedded PC of any description, you will still have some learning curves to climb as far as designing software for this closed-box environment. At the very least, you probably want to have a watchdog timer (WDT) in your box to prevent lockups (especially if you're using an operating system that isn't specifically intended for embedded use). Most single-board computers already have a hardware watchdog on-board, but if you're using a consumer-grade motherboard you'll probably have to build your own. The simplest sort of watchdog is just a dead man's switch that expects to be poked periodically—perhaps by receiving a character over a serial port, or perhaps by regular toggling of a status line. If it doesn't get the required input for some period of time, the watchdog interrupts power for a few seconds and thereby hard-resets the system. More advanced WDTs—available as plug-in cards for common buses such as PCI, and usually integrated directly on industrial single-board computers—listen for special writes to an I/O port, or port range[4] and assert the master system reset line for a period of time if the timeout

---

[4]   One vendor of watchdog cards for PC applications is Berkshire Products, Inc., at <*http://www. berkprod.com/*>. Their PCI watchdog card provides all kinds of goodies including an A/D input, DIP-switch and/or software-configurable WDT timeout, eight digital I/O ports, two relays, temperature trip points, and various other features.

fires. These products usually incorporate other handy system health functions as well, such as temperature zone monitoring. In some cases, you might want to implement your own custom WDT hardware, particularly if you have external mechanical hardware—a sawmill, for instance—that needs to be brought to a safe condition immediately if the software in the control computer is determined to have failed.

Be aware that WDTs on single-board computers tend to have amazingly long timeouts (seconds or minutes) compared to the values you'd find in small microcontroller systems (milliseconds). In some extreme cases, the WDT will be set—in hardware—to fire after as much as five minutes. This long delay is necessary if the WDT is always live; it allows time for a big operating system and application set to load after power-cycling the system. It's a better idea to choose a WDT that is only armed by some explicit software action, and/or has a one-time programmable timeout; that way, you can minimize the worst-case length of downtime after a glitch.

You should also be cognizant of the fact that the sudden-death power cycling approach to watchdog timer behavior is definitely not healthy for some types of hardware. In particular, CompactFlash cards (commonly used in x86 designs based around single-board computers) are *not* designed to handle power inter-ruption elegantly. If you cut power to a CompactFlash card during a write operation, the entire card may be rendered unusable. This situation could easily happen if, for instance, you've got a log file open for writing and you happen to be committing a change at the moment the WDT fires.

You will also need to consider how your application will interface to the real world. In these modern days of legacy-free PCs, it's harder than it used to be to lash home-made peripherals onto the side of the computer. The easiest way to attach digital peripherals such as pushbuttons and relays used to be to wire them to the PC's parallel port. The easiest way to interface a slave controller board to your computer used to be over a serial (RS-232) link. Since both these interfaces are dying out in the consumer PC market, you may need to investigate other methods. Note, by the way, that the SBC market is considerably more embedded-friendly, as you might expect; parallel and serial ports are still standard features on practically all of these boards. However, it might be a wise future-proofing idea to avoid those older interfaces if you can; it will make it easier for you to bring your application over to a different board.

There are various exotic things you can do to attach external custom peripherals to a PC, but the "easiest" route is to go in via USB. (Nothing in USB is truly easy; it's a complicated interface to use. If you're very cautious about what you attempt to do, you can hopefully avoid most of the pain, however.) If you just need a few external pushbuttons and your product is a low-volume device, you can cannibalize a USB mouse or keyboard and wire your own buttons to the appropriate points on the microcontroller in those devices; you won't need to write any device drivers or build any complex hardware. For higher production volumes, or if you need something more complex than just a couple of buttons, more drastic solutions are called for.

Now, I'm a big fan of keeping things as simple as possible, particularly where USB is concerned. In particular, I never want to be in the position of having to write and maintain USB driver code for the host operating system, especially if that host operating system happens to be a rapidly changing consumer operating system like Windows or Mac OS. Therefore, I prefer to use a USB interface solution that has fully debugged, vendor-provided drivers.

From a practical design standpoint, this means that I still use regular RS-232-style asynchronous serial as my external interface of choice. In between the PC and the external hardware, I use USB-to-serial converter chips from FTDI *<http://www.ftdichip.com/>*. These devices are a boon to embedded developers, because FTDI provides completely free drivers for Windows, Windows CE and Mac OS. (Linux has had FTDI support built into the kernel for a while; if you're using an older kernel, FTDI provides a driver for your use.) All you need to do is plug the device in, install the driver, and you're communicating.

You don't even need to put the FTDI chip on your board; you can use the pre-assembled "ChiPi" cable (UC232R) which is available direct from FTDI for about $20. The only real downside to this is that you'll need to put an RS-232 level matcher on the other end of the cable; if you use the bare FTDI chip then you can avoid shifting from CMOS to RS-232 levels and back again. Of course, you can hack the ChiPi so that it outputs inverted CMOS signal levels, if you so desire.

Unfortunately, some applications can't be implemented using the serial over USB method due to tight timing requirements or simply because they involve large data transfers to or from the external hardware. At this point you need to bite the bullet and build some custom USB hardware to talk to the outside world.

There are basically two routes you could take here; either graft a USB interface onto an existing design (the Philips PDIUSBD12 is an example of one low-cost part you can use for this purpose) or use a microcontroller with built-in USB device-side hardware. In general, I prefer the latter approach, and the specific devices I like are the USB microcontrollers from Cypress.[5] The device I would design in today is the EZ-USB FX1™ (CY7C67413); in the past I've used the older EZ-USB parts (AN21xx).[6]

The FX1 part is a high-speed 8051 microcontroller with a USB interface, I²C interface, 16 Kb of code memory, up to 40 GPIO pins, two DPTRs, and various other goodies. The truly nifty part about this chip is that the code memory is actually RAM. On powerup, the chip can load that RAM area with code from an external I²C EEPROM. If your design doesn't include the EEPROM, the chip will sit around doing nothing until you connect it to a USB port. It will then identify itself to the computer as a generic Cypress device. Cypress provides a driver that will download your custom code to the chip while it is in this state. The driver then simulates a disconnect and reconnect, but this time around it's running your code, so Windows loads a different driver to talk to it. Exactly which driver is loaded depends on how your device identifies itself; it could be a standard Windows driver such as a human interface device or storage class driver, or it could be something completely custom.

The reason this is so exciting is partly because it makes developing and debugging the device side of the code nice and fast, even if you don't have the wherewithal for real in-circuit debugging hardware. The burn-test-pray cycle is accelerated by making the burn phase very fast. A more important fact, however, is that you can reprogram the device simply by updating the host-side driver so it downloads new firmware to the device. This allows you to upgrade the firmware in the field without requiring any special hardware, and without taking any risk that a self-updating micro might lose power during the upgrade process and leave you with a brick.

---

5   Other vendors have microcontrollers with on-chip USB, of course. PICs are quite popular for this type of application, too, though they don't have the nifty RAM-based architecture Cypress offers.

6   The reason I mention this older part at all is that DeVaSys *<http://www.devasys.com/>* still offers a relatively inexpensive ($79) evaluation board with the AN2131QC chip on it and all the I/Os brought out to a handy header. This is the cheapest way to get up and running with the EZ-USB parts. I couldn't recommend you use the AN21xx family in a new design, since it is approaching the end of its life; but you can certainly use the part to get a lot of good experience developing embedded USB code.

Unfortunately, regardless of what path you take, if things come to a pass where you're designing your own USB microcontroller code, you'll almost certainly have to write a custom driver on the PC side as well. If you're looking for a relatively low-stress introduction to USB from the embedded developer's perspective, an excellent book is *USB Complete: Everything You Need to Develop Custom USB Peripherals* (3rd edition) by Jan Axelson (Lakeview Research, August 2005, ISBN 1-9314-4802-7). There are numerous tutorials and other information available online, of course, and the full specifications can be downloaded from the USB Implementers Forum, Inc. *<http://www.usb.org>*.

By the way, if your only interface need is a few analog channels or something of that kind, you don't need to build custom hardware of any sort. Several vendors—National Instruments™ *<http://www.ni.com/>*, Dataq® Instruments *<http://www.dataq.com/>* and IOTech *<http://www.iotech.com>* for example, sell data acquisition kits that can be used as the interface core of an embedded PC-based system. Some of these kits are very inexpensive; Dataq's entry-level 8-channel serial data acquisition device, for instance, is only $25. The equivalent USB product is $50.

The other big decision you'll have to make in your x86 project is what operating system will be used. Later in this chapter I discuss a couple of popular embedded operating systems, but I don't mention Windows, which is the OS you probably think of most readily when x86 is mentioned. The reason for this is that standard consumer Windows is emphatically not an embedded operating system. Also, it is *only* available for the x86 environment, whereas most embedded operating systems are portable to other architectures.

For this section—and only for this section—I'll give you some valid reasons that might argue for building your system around Windows. Some people—particularly at Microsoft—would argue that my list is excessively cautious, but for what it's worth, here are the "should have" criteria that I suggest you look for before deciding to design around Windows:

- You are already an experienced Win32 programmer, and/or you have access to an experienced Win32 programming staff. This criterion might override all others in some cases; if you need to deliver a product quickly, and all your in-house talent is Win32-based, it can make reasonably good sense to use what you have.

- Speed of application development is more important than optimizing the bill-of-materials cost and fine-tuning overall device performance. Note that building your application as a Win32 program means you can test it on a regular PC—and, if necessary, show it to customers—while the hardware team is still arguing the details and assembling prototypes of your hardware platform. It's smoke and mirrors, but it's also the reality of modern marketing campaigns.

- Your application has no significant real-time requirements except, perhaps, embedded multimedia functionality (MP3 audio playback, video playback, and so forth). Windows was not designed to provide guaranteed real-time performance, although it's possible to fake it in some cases.

- A significant proportion of your application is GUI user-interface code. If this is the case, you can take advantage of numerous well-tested GUI libraries and development tools for the Windows environment.

- Your application will never be connected directly to an open network such as the Internet. Consumer Windows is constantly phoning home for updates to protect it against new and improved security threats. It's *very* difficult to lock the operating system down tight enough that these updates aren't necessary (especially if you need to have Microsoft-provided network services of any sort running on the box). On the other hand, it's a bad idea to allow an unattended device to be downloading and installing third-party software updates; you have no idea if one of these automatic updates is either going to require user intervention, or adversely affect the performance of your real application.

- You have no real need to hide the underlying "PC-ness" of the device.

- You need to include functionality that's provided only by proprietary third-party software. A good example of this is support for Flash movies. Although at least one group has made surprisingly good progress at reverse-engineering Flash and providing a third-party playback utility, if you want to advertise Flash capability without getting end-user complaints, you really need to be running the Adobe® plugin or standalone player. Adobe, for all practical purposes, only supports Windows, Mac OS and Windows CE; although there is a Linux player, it's not well-supported and it's certainly not open-source, so you can't recompile it for non-Intel platforms.

4.3 ARM

ARM is a fantastically popular embedded core with a most interesting pedigree. ARM is so popular, in fact, that if you learn this one 32-bit core and no others, you'll likely be employable for a long time to come. The name is an acronym that originally stood for "Acorn RISC Machine" but is now generally accepted to mean "Advanced RISC Machine."

I won't delve into the history of the ARM core too deeply, but it's too fascinating to skip entirely, so here is a quick thumbnail sketch: In the early 1980s, a British company called Acorn Computers was manufacturing a range of 8-bit computers based around the 6502. These machines included the Acorn Atom, British Broadcasting Corporation Microcomputer[7] models A and B, Acorn Electron, and BBC Master. (British and Australian readers, among others, may remember the BBC Micro fondly, as it was widely used in primary and secondary schools in the late 1980s. The machines didn't make it to the United States in any significant number due to RF emissions problems causing difficulty with FCC approval.)

In 1983, Acorn began developing their own proprietary RISC microprocessor as an upward migration path away from the 6502, and in 1987 they released the Acorn Archimedes, a powerful personal computer (designed to replace the BBC Micro), based around this new architecture. The Archimedes and its descendants never saw enormous commercial success, and the RiscPC that followed the Archimedes went out of production last century,[8] but the ARM microprocessor was adopted by Apple® as the core of its Newton PDA product line.

Between then and now, the ARM core has enjoyed sales growth that is nothing short of staggering. Today, ARM cores are realized into silicon by dozens of vendors; they are found inside ASSPs in all sorts of products from PDAs to handheld games to mobile phones to rack-mounted networking gear. Intel even practically abandoned its home-grown i960 RISC product line (the principal applications for which included laser printers and cached disk controllers) in favor

---

[7]   Generally referred to as the "BBC Micro". However, the BBC acronym is owned by the Brown Boveri Corporation, which was the cause of some legal action. The formal name of the computer is therefore the British Broadcasting Corporation Microcomputer.

[8]   There is, however, an ARM-based RISC OS PC still on sale from Castle Technology under the Iyonix brand <http://www.iyonix.com/>. This machine is, however, quite thrillingly expensive—comparable in cost to the evaluation boards for the high-end XScale parts.

of an ARM-based design. This product family is called XScale, and it is based on the StrongARM variant originally developed by Digital Equipment Corporation and passed on to Intel as part of a legal settlement during DEC's breakup. The first XScale parts were basically a process migration of the old StrongARM SA1110 part; modern devices are quite different, with considerably faster clock speeds and more flexible peripherals.

The principal interesting characteristics of the ARM core are:

- small size on the chip die,

- low power requirements (in MIPS/mW terms) and,

- available (as prepackaged intellectual property) as both a hard core and synthesizable soft IP. Hence, if you're willing to pay the appropriate licenses, you can even synthesize an ARM into your own FPGA.

Observe, by the way, that ARM (per se) does not manufacture chips—they simply develop the intellectual property relating to the core. It is a frequent neophyte question to ask "Is ARM available from other vendors?" The answer to this question is that there is no such thing as "an ARM"—there are merely chips that contain an ARM core, and such chips are indeed available from numerous vendors. There are no drop-in-replacement, generic multisourced ARM parts in the same sense that there are generic 8051 parts. In fact, as a general rule, many of the ARM parts we see as general-purpose devices now were at one time application-specific products custom-designed for one customer or one particular market space. After such parts have been in production for some time, the chip vendor starts to look for other customers, and if the demand seems to be present, the device will be offered to the public.

The earliest ARM-cored parts (from companies such as VLSI) were microprocessors, rather than microcontrollers. They did not generally have on-chip RAM (except perhaps for caches) or ROM; it was not possible to build a single-chip system using these devices. It was not until about 2003 or thereabouts that we began to see commercially fielded, generally available single-chip ARM products. Possibly the first such product family to the general retail market was the LPC2xxx series from Philips. As of 2006, there are several other vendors with single-chip products; Atmel and ST would appear to be the most popular, though there are other vendors in the field as well. These single-chip parts are very inexpensive, and they are aimed very clearly at usurping the upper end of the 8-bit market. A significant number of high-end 8-bit applications are moving up to ARMs, with a substantial jump in performance for very little additional bill-of-materials cost. For more information on this topic, refer to Section 3.1 and the end of Section 3.2.

To the best of my knowledge, there is no ARM-cored general-purpose part presently manufactured in a package that lends itself to hand-prototyping without a custom PCB. Hence, there are essentially four routes to building a system around these devices:

1. Buy the chip vendor's evaluation board and do your initial development on that. This is usually quite expensive.

2. Buy a third-party evaluation platform for the device. This is usually much cheaper than the first option, but you might be somewhat restricted by whatever other hardware the third-party vendor chose to put on the board.

3. Repurpose an off-the-shelf device that happens to use the chip of interest. As a somewhat relevant example, a lot of people have built interesting things out of the Linksys® WRT54G (now WRT54GL) wireless router—for example, see the information page at: *<http://www.seattlewireless.net/index.cgi/LinksysWrt54g>* and the homepage of Sveasoft at *<http://www.sveasoft.com/>*, where you can obtain a commercial set of replacement firmware for this device (among others). This particular device happens to be based on a MIPS processor, not ARM, but the same principle applies. The problem is, of course, that consumer hardware is a moving target; manufacturers often change the innards of their appliances without changing the model number or packaging—so you can't be assured of a continued supply of whatever version you've chosen to hack.

4. Develop your own circuit from scratch and build your own custom board. This is, needless to say, a high-risk path if you're not already experienced with the particular chip you've chosen to use. I would not go down this path unless your desired application can be achieved by a fairly simple modification of a known-good reference application schematic. (Don't assume that the schematics on a vendor's website are accurate, by the way—I have

almost *never* encountered a vendor's schematic that exactly matched the evaluation board; nearly all of them contain errata that have been patched on the real board but not in the downloadable documentation!)

For the remainder of this chapter, I'm going to discuss the Sharp® BlueStreak LH79520, a representative midrange ARM720T part intended for applications such as low-end PDAs, GPS devices and so on. You should, however, be aware that the ARM family covers a great deal more territory than just this one core; for more information, consult the official reference documentation for ARM's various cores at *<http://www.arm.com/documentation/ARMProcessor_Cores/index.html>*. I'm going to start by talking about under-the-hood details, and then move on to some higher-level application-layer material. Please note that this description barely scratches the surface of what's in an ARM part; it is the material of an entire book to discuss this processor architecture.

The Sharp LH79520 is a 176-pin LQFP device containing an ARM720T core (more on this later) and the following major peripherals:

- Flash/SRAM controller.

- SDR (single data rate) SDRAM controller.

- DMA controller supporting the on-chip LCD controller as well as external DMA devices.

- Vectored Interrupt Controller (VIC).

- Color, bitmapped LCD controller. Note that this controller is intended to drive STN and TFT panels with integral row/column drivers; it is not designed to drive the LCD segments directly.

- 32K of on-chip SRAM.

- Synchronous serial port intended for peripherals such as audio codecs.

- Two PWM outputs.

- Three UARTs, one of which has infrared decode capability.

- Four on-chip timers. The output from one of these timers is routed to an external pin.

- Up to 64 pins of GPIO, multiplexed with the peripherals as previously described.

- A watchdog timer.

- A real-time clock (RTC) module with alarm function. The RTC is clocked by a separate 32.768 kHz watch crystal, and it continues to run while the CPU core is asleep; the alarm function can be used to wake up the core.

- JTAG debugging interface.

Sharp does have an official evaluation board for this microcontroller; however, it's rather expensive (several thousand dollars). There are less expensive solutions available in the third-party market, including the LH79520 Card Engine from Logic Product Development *<http://www.logicpd.com/>* and the mARMalade from EarthLCD *<http://www.earthlcd.com/>*. The latter product is a rather complex but relatively inexpensive single-board computer with Ethernet, a touch screen interface, serial ports, an LCD interface (the board is available in a kit with a color LCD module), a CompactFlash slot and various other goodies. It has the blob bootloader and Linux preinstalled, so you don't even need to fiddle with JTAG adapters and so forth to get the board working; you can do everything over a serial connection.

Most ARM devices, including the LH79520, have several selectable physical memory maps. This feature isn't absolutely necessary on parts, like the LH79520, which have a memory-management unit but, nevertheless, the chip has three memory mapping modes (shown in the following table) selected by the REMAP bits of the RCPCRemapCtrl register.

Range	0b00 (default) or 0b11	0b01	0b10
0xFFFF0000–0xFFFFFFFF	AHB peripherals	AHB peripherals	AHB peripherals
0xFFFC0000–0xFFFEFFFF	APB peripherals	APB peripherals	APB peripherals
0x80000000–0xFFFBFFFF	Not implemented	Not implemented	Not implemented

Range	0b00 (default) or 0b11	0b01	0b10
0x60000000–0x7FFFFFFF	Internal SRAM (repeated)	Internal SRAM (repeated)	Internal SRAM (repeated)
0x40000000–0x5FFFFFFF	External static memory	External static memory	External static memory
0x20000000–0x3FFFFFFF	SDRAM	SDRAM	SDRAM
0x00000000–0x1FFFFFFF	External static memory	SDRAM	Internal SRAM

Note: AHB is the Advanced High-Speed Bus, used for "important" peripherals; APB is the slower Advanced Peripheral Bus, used for peripherals with lower bandwidth such as the UARTs.

The only substantive difference between these modes is what lies at the bottom of the physical memory map. At power-on reset, the ARM core begins execution at location 0x00000000. Hence, the POR default setting for the remap register is to put external static memory (which would normally be Flash or EPROM) in this location. For the LH79520, the default is to assume rather slow 16-bit memory from 0x00000000 to 0x04000000. Actually, the chip has seven static memory banks, each of which has a dedicated chip select line coming out of the micro.

Each bank is uniquely configurable as to width and speed, except that in order to meet the LH79520's boot requirements, you'll want to have 16-bit Flash or other nonvolatile memory connected to chip select 0. The default bank assignments are shown in the next table (note that these addresses are offsets from the start of the static memory area; add 0x40000000 if you want to access the "permanent" location of static memory, or use these addresses directly if you're in remap mode 0b00 or 0b11):

Range	Width
0x00000000–0x03FFFFFF	16-bit
0x04000000–0x07FFFFFF	16-bit
0x08000000–0x0BFFFFFF	16-bit

Range	Width
0x0C000000–0x0FFFFFFF	8-bit
0x10000000–0x13FFFFFF	32-bit
0x14000000–17FFFFFF	32-bit
0x18000000–0x1BFFFFFF	16-bit

The normal sequence of events at powerup would be as follows:

- A small bootstrap program at 0x00000000 (in physical memory) jumps to a location in the 0x40000000–0x5FFFFFFF range. Note that this is still executing out of the same physical memory device, since those two locations are mirrored; it's merely a change in the program counter.

- The remainder of the bootstrap program initializes static memory controller and SDRAM controller to match the product's physical memory map and the attached devices. Note, by the way, that memory-mapped peripherals are attached to the static memory controller.

- The main program image is copied to SDRAM or on-chip RAM.

- The remap register is set to either 0b01 or 0b10 (0b01 would generally be more usual).

- The bootstrap program jumps to location 0x00000000, which is now pointing at high-speed RAM, usually 32 bits wide.

- Main program image execution begins. Observe, by the way, that the ARM vector area at 0x00000000 is now in RAM.

Now, an important point: The ARM core operates with 32-bit words. It expects to have external hardware manage the byte lanes for unaligned accesses and memories that are less than 32 bits wide. Therefore, it is absolutely critical that you take care to program the static memory controller for the correct width of the device you're accessing, and (particularly in the case of memory-mapped I/O devices) that you use the correct instructions to access memories that are narrower than 32 bits. If you don't heed that last warning, you'll experience some very odd behavior, because the static memory controller (SMC) will select the wrong byte lane(s) and may generate additional, unwanted read and write cycles. For instance, if you have a 16-bit peripheral on the bus, and you configured the SMC correctly, a 32-bit read or write will cause the SMC to issue two read or write cycles to the lower 16 data bits. Conversely, if you have an 8-bit peripheral on the bus, and you accidentally configured the SMC for 32-bit operation on that bank, reads and writes for addresses that are not an integer multiple of 4 will be routed to the wrong data bus lines.

While we're on the topic of memory, a brief word about the memory management unit (MMU). The ARM720T has an 8K on-chip cache memory area. In order to use this, you must enable the MMU, since the cache enable bits are set on a per-page basis in the MMU page tables. The MMU operates with two data structures; a table of "section" entries that describe memory in 1 MB blocks, and (optionally) one or more tables of "page" entries that describe memory with finer granularity, in either 4 KB or 64 MB blocks. The MMU translation table must reside on a 16 KB boundary in memory.

Each entry in either of these tables is a 32-bit word. You'll find the address translation process, and how to create the page tables, described in detail in section 6 of the *ARM720T Rev 3 Technical Reference Manual*, but in brief: When the processor attempts to read or write a given address, the address is divided by 1 MB and the result is used as an offset into the level 1 (section) descriptor table. The 32-bit entry thus fetched either describes fully the handling to be given to this 1 MB zone, or it references a page table (level 2 descriptor). Bits in the table entry control whether the area is cached and/or write-buffered, which privilege domain the area belongs to, and where it should be mapped in physical memory (with 1 MB granularity in the case of the section table). The simplest way to deal with the MMU is to create a set of section entries that maps all of your memory into a contiguous area starting at 0x00000000, with cache and buffer enabled where appropriate. (Since the size of the table is directly dependent on how large your addresses get, cramming everything down to the bottom of the section table minimizes the number of entries the table needs to create.) Advanced operating systems, particularly those that implement virtual memory, will need to implement page tables so they can have finer-grained control over memory permissions. As with practically all MMUs, the ARM720T's contains a translation lookaside buffer (TLB) that caches 64 sections' worth of mappings, so they don't need to be looked up.

From an assembly language programmer's standpoint, the ARM720T is a von Neumann, highly orthogonal RISC architecture with a 32-bit instruction word. The "T" suffix indicates that this part also supports the Thumb mode. In Thumb mode, the instruction word is only 16 bits in size, offering an improvement in code density and potentially an improvement in speed if the code is being executed out of a 16-bit memory device. However, while Thumb mode is active, not all the general-purpose registers can be accessed, some other functionality is unavailable, and algorithm execution is therefore slower, all other things being equal. For the remainder of this discussion, unless otherwise specified, I am talking about ARM mode (the 32-bit mode), not 16-bit Thumb mode. Note that all instructions in ARM mode can be given a conditional override; this is not available in Thumb mode.

The ARM core has sixteen general-purpose registers named r0 through r15 (in Thumb mode, r8 through r12 are not directly accessible). R15 is reserved for use as the program counter (PC), r14 is used as a link register for branch instructions (ARM assemblers give it the alias "lr"), and r13 is the stack pointer (SP). There is also a status register, CPSR (current program status register). This status register is active for conditional branches and so forth while the processor is executing in the unprivileged "user" and "system" mode. When the core is kicked into the privileged modes—FIQ, Supervisor, Abort, IRQ or Undefined—dedicated SPSR, LR and SP registers are mapped in, so that each of these modes can have its own link register, stack pointer and status word (these don't need to be saved when you exit the privileged mode; they are kept separately in the core).

If you're not familiar with high-performance RISC architectures, you might find the idea of the link register a bit alien, by the way. CISC devices typically execute a call instruction by pushing the current program counter onto the stack and jumping to the new destination address; the matching return instruction automatically pops the return address off the stack into the program counter. RISC devices, on the other hand, typically have a link register and a special form of branch instruction that is usually called *branch with link*, or something similar (the mnemonic in ARM assembly language is "bl"). The branch with link instruction operates by copying the return address into the link register—however that is defined in the processor's instruction set architecture—and jumping to the destination address. To return to the called program, you use an instruction that copies the link register back into the program counter; on the ARM, this would be either "bx lr" (unconditional branch to link register) or "mov pc, lr" (copy link register to program counter)—note that the two are not quite identical.

The advantage of this system is that you can avoid touching memory when making nonnested calls inside the inner loop of a function; it can be a significant performance enhancement. The called function also doesn't *necessarily* need to mess around with stack arithmetic if it created some local variables on the stack; it can jump directly back to the caller using the link register, the caller can simply restore the stack pointer to what it was before the function call, and all local variables and passed parameters would be cleaned up automatically. (The usual ABI for ARM specifies that r0–r3 are used to pass the first four parameters to C functions; any remaining parameters are passed on the stack.) You can think of the link register as, conceptually, a high-speed cache of the last return address on the stack. It only needs to be written out to actual RAM when making a nested function call. Additionally, when the link register is not being used as such, it can be employed as a general-purpose register.

I've spent some time talking about how nifty ARM is; so, what are the downsides? Not many, really. There are three principal downsides. First, the interrupt architecture is a bit unusual—the core only directly supports two interrupt modes, IRQ and FIQ. IRQ (Interrupt ReQuest) is the "lower priority" interrupt; FIQ (Fast Interrupt reQuest) is a high-priority interrupt that doesn't need to save quite so much processor state, but guarantees lower latency than IRQ. FIQ is supported by shadow registers covering r8–r14 and SPSR (the status word); you don't need to save these registers during the interrupt, and as a result your latency is reduced. Since there is effectively only one interrupt level in the core, interrupts from multiple sources have to be vectored using software techniques; this is accomplished by effectively jumping into the register contents of the Vectored Interrupt Controller. It works, but it's a bit strange.

Second, on a closely related note, interrupt latency on ARM is relatively poor (potentially tens of clock cycles) compared with the 8-bit parts we all know and love. When an interrupt occurs, the CPU has to complete the current instruction—this could be as much as 15 clock cycles for a multiword store instruction to 32-bit memory in ARM mode (potentially worse for narrow memory situations). The interrupt also causes a mode switch, which requires finite time—and the pipeline is flushed, which delays completion of the first instruction in the ISR.

Finally, ARM doesn't handle bitwise data access very efficiently compared with some 8-bit cores (8051, for instance). This makes implementing fast proprietary serial protocols and similar algorithms (decoding Manchester data, for example) less efficient in ARM than on those 8-bit cores.

Both hardware and software tool support on ARM are very good, as you'd expect for such a popular core. Practically all modern ARM parts have on-chip JTAG debugging capability, and many vendors have low-cost wiggler type JTAG adapters. The cheapest of these used to be Macraigor's Wiggler, but an even less expensive version is now available from Olimex. They don't specifically call out support for parts like the BlueStreak, since Olimex doesn't make any evaluation boards around this family of chips, but the ARM core is more or less generic, so if it works with one, it should work with most of them.

Speaking of software, most of the architectural stuff I've discussed previously is really of greatest interest to the low-level guys working on system startup code and perhaps some timing-critical device driver code. In many ARM environments, you'll be working with a bootloader—blob *<http://sourceforge.net/projects/blob/>* is one such bootloader,[9] and another popular candidate is U-Boot *<http://sourceforge.net/projects/u-boot>*. Still another choice might be RedBoot *<http://sourceware.org/redboot/>*. All of these programs have the same basic functionality: they load the operating system image and jump into it. In addition, the bootloader provides a user interface over a serial port (or perhaps a telnet interface) allowing you to override the normal boot process and load new operating system images onto the device's internal storage device, whatever that may be. You can also select among different boot images on the device, if more than one is offered—in some cases, you can also select to have the device look on the network for a tftp boot server and load its operating system off that, if possible.

For the majority of ARM applications, your bootloader will configure the initial system environment and load your operating system image from secondary storage (this might be a location in boot Flash, or something in NAND Flash, or an image on a hard drive or even over a network). Hence, by the time your code gets to run, most of the boot-time details I previously discussed are taken

---

9   Note that the website describes blob as a StrongARM bootloader—in fact, it is used on other ARM parts as well.

care of already—the most you might have to do is set up the memory management unit.

As far as software tools go, you have several options: Microsoft makes a set of software tools for Windows CE development support, Keil makes their own ARM C compiler (you can expect this to be the best game in town, since ARM actually owns Keil). Rowley Associates also has a good package, although at the time of writing their product is simply a frontend and nice IDE wrapper around the standard GNU toolchain. As with the other cores I've discussed so far, my preference is to use the GNU tools for this part. Probably the best support for "raw" ARM programming is available from *<http://www.gnuarm.com/>*—here you will find prebuilt toolchains for Cygwin (Windows), Linux and Mac OS. Another prebuilt toolchain is WinARM, *<http://www.siwawi.arubi.uni-kl.de/ avr_projects/arm_projects/#winarm>*, and still another is provided directly by Macraigor at *<http://www.macraigor.com/>*.

The reason I said "raw" programming there is that the commonly available prebuilt toolchains are linked with newlib, the embedded C runtime library. If you're working with an operating system, you probably want a toolchain built with the correct startup code and runtime library for your operating system. For example, if you're working with ARM Linux, you want to visit *<http:// www.arm.linux.org.uk/>* for more information on the correct cross-compiling toolchain.

Note that you are certainly not restricted to using C on these powerful micros; C++, Java and numerous other languages are directly supported by the GNU tools and others. (Some ARM variants—not the LH79520, however—even have hardware acceleration for Java bytecode interpretation. This coprocessor is called *Jazelle*.) The language you choose for your application layer will depend to a large degree on what you're trying to do.

Shameless advertisement: If you're looking for an introduction on how to get started with the ARM processor core and GNU tools, you might want to check out my first book, *Embedded System Design on a Shoestring* (Newnes, 2003, ISBN 0-7506-7609-4). This book will guide you through building the GNU ARM tools in Windows or Linux (the same instructions should work for Mac OS, although I haven't actually tested it). It will also show you how to write makefiles, and guide you step by step through creating the startup code necessary to get a working C environment on a small ARM processor. The book is targeted at an older evaluation board, the Atmel AT91EB40, but the principles can be migrated to almost any evaluation platform. This book is intended to help you get up and running on a bare board with no operating system at all; you might not find it very useful if you're starting with an ARM platform that has, say, Linux or Windows CE preloaded.

Speaking of this, I should point out that, obviously, proprietary home-rolled operating systems and Linux are certainly not your only choices. Many operating systems are available for ARM parts; Symbian (used in many cellphones), eCos, Windows CE, the Palm OS®, VxWorks, uCos-II, NetBSD, . . . the list is very long. Later in this section, I'll give you a brief introduction to eCos and Linux in the context of embedded products. The reason I suggest these is that they're good learning tools (as well as being perfectly viable tools in their own right), they're open-source and they're entirely royalty-free. If you're preparing yourself for an entry into professional embedded development, a good grounding in these two operating systems will prepare you to work with other operating systems at a detailed level; of course, there are plenty of jobs working on embedded Linux outright, too. (eCos isn't used as widely, but it is out there—you just won't see it in a job posting very often.)

If you buy a pre-built ARM single-board computer of any type, it is likely to come with your choice of either ARM-Linux or Windows CE. In order to get up and running with the ARM-Linux variant, you are going to have to learn at least a little about Linux's startup process, so that you can at least get your application installed and launched; see Section 4.5 for more detail on this. However, you will probably not need (at least initially) to learn how to rebuild and install a new kernel, or how to set up an entire Linux root filesystem from scratch.

4.4   PowerPC

The PowerPC family, manufactured by IBM® and Freescale, is a high-end 32- or 64-bit platform used primarily in applications where performance is the most important factor. The latter half of that statement is, of course, a generalization, but it fairly accurately reflects the situations where you're likely to find PowerPC. Historically, the part originated from an experimental IBM chipset created as essentially a research project during the late 1970s. This eventually metamorphosed into the POWER architecture used in the RS/6000 range of UNIX workstations and servers. RS/6000 has been through some nomenclature changes, and the current incarnation of these machines is known as the eServer p5—it runs both AIX and Linux. The PowerPC architecture is derived from POWER; you can read a very detailed and quite fascinating history of the parts at *<http://www-128.ibm.com/developerworks/power/library/pa-powerppl/>*. The most recent step in the evolution of the PowerPC family is the immensely powerful Cell Broadband Engine used in the new PlayStation console; this chip surrounds a high-performance PowerPC core with several coprocessors, all on the one die. It's hard to overstate just how amazing this chip is; it can perform highly complicated graphical transforms, process an incoming compressed digital video stream in real time, and run complex artificial intelligence algorithms controlling your onscreen enemies.

PowerPC's consumer visibility suffered something of a blow in 2005, as Apple, one the highest-profile customers of PowerPC devices, announced plans to migrate their computers to Intel x86 cores.[10] This event, however newsworthy, is more or less irrelevant to embedded developers; most of the PowerPC devices sold are not part of general-purpose desktop computers, but are rather embedded processors used in communications, telematics and other applications. As an interesting data point, it should be noted that PowerPC is used in all the high-end video game consoles currently on the market; these applications swamp Apple's output of personal computers by a couple of orders of magnitude.

As an interesting counterpoint to Apple's migration from PowerPC to x86, in 2005 I began a series of articles for IBM's developerWorks resource, describing how to migrate a complex embedded application the other way—from x86 to PowerPC. A little while later, I began another series describing how to build a multimedia appliance around a PowerPC (having previously been in the industry of building similar appliances around the National Semiconductor Geode platform).

---

[10] It's the prerogative of authors and prophets to make public predictions that are either shockingly true or hilariously wrong. Here's mine: Apple's decision to move to the x86 core, and particularly the release of Boot Camp (a utility that allows you to run Windows on Intel-based Apple machines) has capped the lifespan of Mac OS at somewhere in the region of five to ten years. It remains to be seen if Apple can continue to sell boutique-priced computers once they become just another Windows OEM. Of course, by the time this happens, they may already be selling nothing but iPods and similar appliances.

You can find links to all these articles at my index page, <*http://www.larwe.com/technical/current.html*>; they are freely accessible on the developerWorks site, and no registration is required to view them.

Getting started with embedded PowerPC development is rather more expensive than working with most of the other architectures mentioned in this chapter. The reason for this is that it's a fairly costly and complex proposition to design your own PowerPC-based hardware platform; it's much easier to start with an off-the-shelf reference board. Unfortunately, the pricing of PowerPC boards is, in general significantly more expensive than for the other architectures we've discussed here; it seems that not many people make really generic single-board computers based around these parts, and the chip vendors' evaluation boards are quite costly.

In the article series previously mentioned, I discussed using a product called the Kuro Box as your development platform. This is a Japanese-made Network Attached Storage (NAS) device with an MPC8241 processor (a "G2" core, if you're familiar with that nomenclature), an internal hard disk (not actually supplied with the Kuro Box unit; you have to provide your own), Ethernet, USB 2.0, 4 Mb Flash and 64 Mb RAM, running Linux. It's a slightly unusual product in that it was originally a consumer device—the Buffalo Linkstation, to be precise—until consumers started hacking it to do other things. Buffalo thought about this for a while, then decided to release the device in special packaging, specifically aimed at highly technical people who aren't afraid of Linux and want to build a custom appliance attached to their LAN. This device is the cheapest way to get into PowerPC development at the moment; it's available in the United States from Revolution, <*http://www.revogear.com/*>.

Another vendor of PPC hardware you might want to consider is Genesi USA, Inc. This company makes several PPC workstation devices under the Pegasos brand name; you can see their product range at <*http://www.pegasosppc.com/*>. (Interestingly, Genesi used to advertise a forthcoming laptop called the *4U2*, based around the MPC5200 chip and closely related to the EFIKA 5K2 design. The 4U2 has, however, disappeared off the roadmap.) The Genesi devices are unfortunately quite expensive for what they contain; simply buying a PowerPC Apple Macintosh® is still by far the cheapest way to get into PowerPC development, particularly if you're developing at the high end.

I'd suggest that the best compiler to use for PowerPC is gcc—end of story. The reason for this is that IBM is heavily invested in Linux and open source technologies in general, and they are providing significant input into the compiler development as well as the operating system. The only two PowerPC operating systems with a relatively large installed consumer base—Linux and Mac OS—both use gcc as their build environment. There is, however, an alternative in the form of Altium's TASKING C/C++ compiler and debugging environment; for more information, view *<http://www.tasking.com/products/32_bit/ppc/>*. This compiler is only supported for use with the CMX-RTX and RTXC operating systems, however.

Speaking of compilers, it's not likely you will spend much time working with PowerPC assembly language. The instruction set architecture is sufficiently complex that it's definitely best optimized by your compiler. Even the relatively simple G2 core in the MPC8241 is strikingly difficult to understand. It has 32 general-purpose registers (GPR0-31) and 32 floating-point registers (FPR0-31), a condition register (CR), floating point control register (FPSCR), link register (LR), and almost a thousand *documented* special-purpose registers. And that's just the core—not counting the peripherals in the rest of the chip! It's very difficult to wring out better performance than you'll see from a good C compiler; you need to take account of pipeline state, cache behavior, branch prediction (each conditional branch instruction has a "hint" in it that tells the core which way you think the branch is more likely to go—the pipeline will start filling from that destination). Of course, it's possible to program this part in assembly language—and necessary for low-level developers to at least read the assembly language fairly fluently—but there is simply too much to learn about the core for me to present a meaningful insight into it for you here, so I'll refer you to the *G2 PowerPC Core Reference Manual*, Freescale document #G2CORERM/D, which is downloadable from the product information page for the MPC8241 chip at *<http://www.freescale.com/>*.

Operating systems you are likely to consider for homebrewed PowerPC systems include Linux (the homepage for the PowerPC variant is *<http://www.penguinppc.org/>*), NetBSD, VxWorks and similar UNIX-based offerings. Microsoft used to offer a version of Windows NT for PowerPC systems, but support for this was dropped (and in any case, it wasn't intended for embedded systems). There are some more specialized operating systems available (such as the aforementioned CMX-RTX and RTXC) but these aren't really products for which you'd just buy the CD and start coding. Similarly as for ARM, you'll probably buy your PPC platform with an operating system support package already ported and ready to run; you can start building your application on top of that without getting into too many gory details from the start.

Working with PPC at a high-level application level is, frankly, almost indistinguishable from working with x86, a fact that I demonstrate in some detail in my Kuro Box article series. The interfaces you'll use—PCI, USB and so forth—work much the same way, and in fact oftentimes you'll have regular PC peripheral chips connected to a PowerPC processor, so you can use existing driver code verbatim (or nearly so). The only concrete difference you're likely to see is better MIPS/mW numbers and, in some cases, better clock-for-clock performance on tasks such as software DSP algorithms. Unless you're doing something particularly hairy, porting (say) a Linux program from x86 to PowerPC is simply a matter of recompiling. This is particularly handy for the small development team on a tight budget, because it means you can develop and demonstrate your application entirely on a regular desktop PC running Linux, and rebuild it on the real hardware once it's ready for primetime.

4.5 Linux

Almost everyone who would be reading this book knows at least some of the history of Linux and how it was designed by Linus Torvalds, so I won't dig deeply into that topic—if you want more information, the Wikipedia entry for Linux *<http://en.wikipedia.org/wiki/Linux>* is probably as good a place as any to learn about it. Linux is unarguably the world's most popular open-source operating system. It's available for numerous platforms, and in its current incarnations includes (where supported by hardware) a complete turnkey GUI windowing environment, XFree86. Linux is used on desktops, in Web servers, TiVo digital video recorders, and in PDAs, among many other things.

Architecturally, Linux consists of a modular, multitasking kernel (this is the actual operating system per se) and numerous support programs such as the command line shell. Most of the time you're working with Linux, you're interacting with these support programs, not the actual operating system, so some people get confused and think that, for instance, the shell is Linux.

The basic Linux kernel can be downloaded from *<http://www.kernel.org/>*. Some architecture-specific ports also exist, such as ucLinux (a special version of the OS for microcontrollers that lack a memory-management unit), arm-linux (Linux with patches for various ARM-based platforms), and others. In order to build any of these kernel source distributions into an actual Linux system, you need a considerable amount of support magic, ranging from a bootloader to get the kernel into RAM off secondary storage, to an init program that will manage system startup and shutdown. Also, of course, you need your own application, which is just another program as far as Linux is concerned (even though it may consume 100% of your system's resources!).

The fantastic thing about Linux is the vast array of hardware support in the kernel and various add-on packages such as XFree86 (the GUI windowing system). This is a great boon to the embedded developer, because you can immediately get up and running with all sorts of heterogeneous hardware without having the tedium of writing your own drivers.

Linux is licensed under the GNU General Public License (GPL), which fact is the cause of much unnecessary angst—mostly due to falsehoods and half-truths disseminated by vendors of proprietary software. Simply put, the GPL requires that if you modify a GPL'd piece of software and distribute it to anyone, your code is covered by the GPL also and you must supply the sourcecode to your modifications on request. However, there is an exception, which is that if your program only accesses GPL'd code through documented interfaces, it does not need to be GPL'd itself. What this means, as a practical matter, is that you will need to disclose sourcecode if you modify the kernel or any of the other GPL'd programs that form part of the typical Linux install, but if you simply write a Linux application, you won't be pulled into that requirement. Speaking of GNU, by the way, note that the only "safe" tool to use for building Linux is the gcc toolchain.

Linux has been embraced by both vendors and customers in the embedded arena; there are literally hundreds of ports, and almost any 32-bit embedded platform you look at these days will probably come with Linux as a build option. A happy side-effect of this is that "porting" Linux to a new platform is often no work at all—if the microcontroller is supported at all, chances are that all you'll have to do is set the right kernel options, possibly include some additional drivers for hardware that you've put on your board, and build it.

In the remainder of this section, I'll introduce you to some of the details of the Linux startup process to illustrate what's involved in embedding Linux. Most of this text was first published on IBM's developerWorks website in an article I wrote in 2005; it has been edited for coherence with this book. If you're interested in Linux development on PowerPC, I suggest you read the entire series of articles, which you can access through an index page I maintain: *<http://www.larwe.com/technical/current.html>*.

By the time a system has booted to the point where it can run your application-level code, any one variant of Linux is, practically by definition, largely similar to another. However, there are several different methodologies that you can use to get the system from power-on reset to a running kernel, and beyond that point, you can construct the filesystem in which your application will run in different ways.

Each approach has its own distinct advantages and disadvantages, and a definite, two-way relationship exists between the hardware you choose to implement and the way you will structure the power-up and Initial Program Load (IPL) process. Understanding the software options available to you is a critical part of the research you must do before designing or selecting hardware.

The most fundamental and obvious difference between x86 boards and embedded systems based on PPC, ARM, and others is that the x86 board will ship with one or more layers of manufacturer-supplied "black box" firmware that helps you with power-on initialization and the task of loading the operating system out of secondary storage. This firmware takes the system from a cold start to a known, friendly software environment ready to run your operating system. Figure 4.1 is a diagram of the typical PC boot process, with considerably more detail than you tend to find in PC-centric literature.

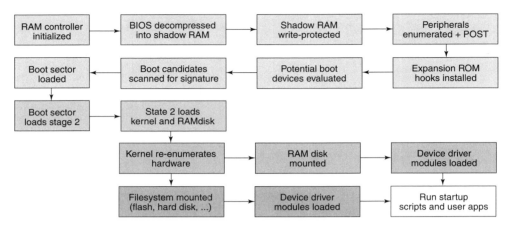

Figure 4.1 A diagram of the typical PC boot process.

For cost reasons, modern PC mainboard BIOS code is always stored compressed in Flash. The only directly executable code in that chip is a tiny boot stub. Therefore, the first task on power-up is to initialize the mainboard chipset enough to get the DRAM controller working so that the main BIOS code can be decompressed out of Flash into a mirror area in RAM, referred to as shadow RAM. This area is then write-protected and control is passed to the RAM-resident code. Shadow RAM is permanently stolen by the mainboard chipset; it cannot later be reclaimed by the operating system. For legacy reasons, special hardware mappings are set up so that the shadow RAM areas appear in the CPU's real-mode memory map at the locations where old operating systems like MS-DOS would expect to find them.

Keep in mind that the PC is an open architecture. This openness even extends down to firmware modules within the BIOS. Once the power-on initialization (POI) code has run, the next step it takes is to enumerate peripherals, and optionally install hooks provided by expansion ROMs in those peripherals. (Some of those expansion ROMs—for instance, the video BIOS in a system that has onboard integrated video hardware—will physically reside in the main BIOS image, but conceptually they are separate entities.) The following are the reasons the BIOS has to do this redundant initialization.

1. The main BIOS needs basic console services to announce messages and allow the user to override default start-up behavior and configure system-specific parameters.

2. Historical issues limit the size of a user-supplied bootloader program to slightly less than 512 bytes. Since this isn't enough space to implement all the possible device drivers that might be required to access different displays and storage devices, it's necessary for the BIOS to install standardized software interfaces for all installed, recognized hardware that might be required by the bootloader.

Once all the BIOS-supported system peripherals are initialized, the main BIOS code will run through candidate boot devices (in accordance with a user-configurable preference list) looking for a magic signature word. Storage devices for IBM®-compatible PCs have historically used a sector size of 512 bytes, and therefore the BIOS only loads the first 512 bytes from the selected boot device. The operating system's installation program is responsible for storing sufficient code in that zone to bootstrap the remainder of the IPL process.

Although it would be possible to write a minimalist Linux bootloader that would fit into such a space, practical Linux bootloaders for the PC consist of two stages: a small stub that lives in the boot sector, and a larger segment that lives somewhere else on the boot medium, usually inside the partition that contains the root filesystem. LILO and grub are the best-known bootloaders for mainstream Linux installations, and SYSLINUX is a popular choice for embedded distributions.

The primary purpose of the bootloader is to load the operating system kernel from secondary storage into RAM. In a Linux system (x86 or otherwise), the bootloader can also optionally load an initial RAMdisk image. This is a small filesystem that resides entirely in RAM. It contains a minimal set of modules to get the operating system off the ground before mounting the primary root filesystem. The original design purpose for initial RAMdisk support in the kernel was to provide a means whereby numerous optional device drivers could be made available at boot time (potentially drivers that needed to be loaded before the root filesystem could be mounted).

You can get an idea of the original usage scenario for the RAMdisk by considering a bootable Linux installation CD-ROM. The disk needs to contain drivers for many different hardware types, so that it can boot properly on a wide variety of different systems. However, it's desirable to avoid building an enormous kernel with every single option statically linked (partly for memory space reasons, but also to a lesser degree because some drivers "fight" and shouldn't be loaded simultaneously). The solution to this problem is to link the bare minimum of drivers statically in the kernel, and to build all the remaining drivers as separately loadable modules, which are then placed in the RAMdisk. When the unknown target system is booted, the kernel (or start-up script) mounts the RAMdisk, probes the hardware, and loads only those modules appropriate for the system's current configuration.

Having said all that, many embedded Linux applications run entirely out of the initial RAMdisk. As long as you can spare the memory—8 MB is usually more than enough—it's a very attractive way of organizing your system. Generally speaking, this is the boot architecture I favor, for a few reasons:

1. The root filesystem is always writeable. It's much less work to have a writeable root than it is to coerce all your other software to put its temporary files in special locations.

2. There is no danger of exhausting Flash memory erase-modify-write lifetimes or of corrupting the boot copy of the root filesystem, because the system executes entirely out of a volatile RAM copy.

3. It is easy to perform integrity-checking on the root filesystem at boot time. If you calculate a CRC or other check value when you first install the root filesystem, that same value will be valid on all subsequent boots.

4. (Particularly interesting to applications where the root filesystem is stored in Flash.) You can compress the boot copy of the root filesystem, and there is no run time performance hit. Although it's possible to run directly out of a compressed filesystem, there's obviously an overhead every time your software needs to access that filesystem. Compressed filesystems also have other annoyances, such as the inability to report free space accurately (since the estimated free space is a function of the anticipated compression ratio of whatever data you plan to write into that space).

Notice a few other points from Figure 4.1. The first is that the grayscale coding is meaningful. In the light gray boxes, the system is running BIOS code and accessing all system resources through BIOS calls. In the mid-gray boxes, the system is running user-provided code out of RAM, but all resources are still accessed through BIOS calls. In the slightly darker boxes, the system is running

Linux kernel code out of RAM and operating out of a RAM disk. Hardware is accessed through the Linux device driver architecture. The darkest boxes are like the slightly lighter boxes, except that the system is running out of some kind of secondary storage rather than a RAMdisk. The rules being followed in the white box are system-specific.

You'll observe from this that there are two possible boot routes (actually, more) once the kernel has been loaded. You can load an initial RAMdisk and run entirely out of that, you can use the initial RAMdisk and then switch over to a main root filesystem on some other storage medium, or you can skip the initial RAMdisk altogether and simply tell the kernel to mount a secondary storage device as root. Desktop Linux distributions tend to use the latter design model.

Also note that there is an awful lot of redundant code here. The BIOS performs system tests and sets up a fairly complex software environment to make things cozy for operating systems like MS-DOS. The Linux kernel has to duplicate much of the hardware discovery process. As a rule, once the kernel loads, none of the ROM-resident services are used again (although there are some exceptions to this statement), yet you still have to waste a bunch of RAM shadowing that useless BIOS code.

In contrast to the x86's complex boot process, an embedded device based on a PowerPC®, ARM or other embedded processor jumps as directly as possible into the operating system. Although there are extant standards for implementing firmware interfaces (equivalent to the PC ROM-BIOS) in PowerPC® systems, these standards are rarely implemented in embedded appliances. The general firmware construction in such a system (assuming that it is based on Linux) is that the operating system kernel, a minimal filesystem, and a small bootloader all reside in linearly accessible Flash memory.

At power-up, the bootloader initializes the RAM controller and copies the kernel and (usually) the initial RAMdisk into RAM. Flash memory is typically slow and often has a narrower data bus than other memories in the system, so it's practically unheard of to execute the kernel directly out of Flash memory, although it's theoretically possible with an uncompressed kernel image.

Most bootloaders also give the user some kind of recovery interface, whereby the kernel and initial RAMdisk can be reloaded from some external interface if the Flash copies are bad or missing. Off-the-shelf bootloaders used in these

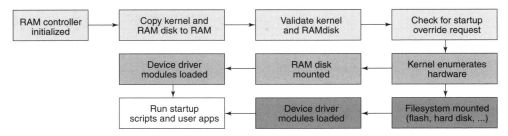

Figure 4.2 A typical start-up flow for a non-x86 embedded Linux device.

applications include blob, U-Boot and RedBoot, although there are others—and there are many applications that use utterly proprietary bootloaders. Figure 4.2 illustrates a typical start-up flow for a non-x86 embedded Linux device.

Observe that, as for the preceding x86 startup process, you have the same possible different routes once the kernel has been loaded. Also note that once control passes to the kernel, the boot process is identical to what it was on the x86. This is to be expected: the further you get in the boot process, the more the software environment is defined by the operating system's API specification rather than the vagaries of the underlying hardware.

The exact layout of such a system in Flash memory depends on two principal factors: the Flash device sector size (usually in the neighborhood of 64 KB), and the processor's power-on-reset behavior. A core like ARM, which starts execution at address 0, will put the bootloader at the bottom of Flash. A core like x86 will need to put the bootloader at the top.

There are at least two, and generally four, entities that need to be installed in Flash: the bootloader (mandatory), an optional parameter block providing nonvolatile storage for boot options, calibration data and other information, the Linux kernel (again, mandatory), and almost always an intial RAMdisk image. For example, a layout for a 4 MB Flash chip with a 64 KB sector size might be as follows:

000000–01FFFF Bootloader (128 KB)

020000–02FFFF Parameter block (64 KB, probably mostly unused)

030000–1FFFFF Kernel (1.8 MB)

200000–3FFFFF Initial RAMdisk image (2 MB)

While it is possible to write these various segments across sector boundaries (and it is especially tempting in the case of the parameter block, which will likely be more than 99% empty), this is an extremely unwise practice and should be avoided unless you are under terribly severe Flash space constraints. It is particularly vital that the bootloader should reside in a private segment that can be left write-protected. Otherwise, a failed firmware upgrade operation may leave the system entirely inoperable. Good system engineering should provide a safe fallback position from any possible user-initiated upgrade process.

The only part of this software bundle that absolutely must be preloaded at the factory is the bootloader. Once the system is startable from that boot code, you can use other (end-user-accessible) interfaces to load the kernel and RAMdisk image.

By the way, at this point the attentive reader may be wondering why embedded PC applications can't use a special boot ROM that simply loads the operating system kernel directly off disk (or some other medium).

The answer to this is that while it's possible to write a custom cut-down bootstrap program for a PC motherboard (see, for example, the LinuxBIOS project), the types of applications that use PC hardware tend to be using the board as a black box. Typically, the system integrator will not even have access to datasheets or schematics for the board; they can't write a bootstrap program even if they want to. Furthermore, PC operating systems are built on the assumption that lowest-common-denominator BIOS services are available, at least at boot time. In other words, it's a simple fact that the path of least resistance is so much easier than a fully custom alternative that practically nobody tries to do it the "smart" way. The inefficiencies of the multilayer BIOS approach are lost in the noise (as it were) compared with the overall system specifications.

Having digested all the previous information, assuming you understand approximately how large your various software modules will be, you are well prepared to select Flash and RAM sizes and layouts for a custom embedded system. Many devices use a very uncomplicated memory architecture; they will have a single linear Flash chip (NOR Flash, i.e., bootable) large enough to hold the bootloader and compressed operating system kernel, and a relatively large area of SRAM or SDRAM, typically between 16–64 MB. While this is the simplest design, it is not necessarily the cheapest, and you may wish to consider other alternatives if you are designing your own system.

One hardware architecture that I have used with some success, and which I have also seen in a few other commercial products, is to use a very small, cheap (generally narrow-bus) OTP EPROM as the primary boot device. This chip is factory-programmed with just enough bootstrap code to load the main firmware image off a secondary storage device and check its integrity. It is very useful if you can also include a little additional intelligence so that the secondary storage device can be reloaded from some external source—removable media, a serial port, Ethernet, USB or something else—if the main firmware image becomes corrupted.

An attractive choice of storage device for the main image is NAND Flash, which is cheaper than the linear NOR Flash used for boot purposes. NAND Flash is produced in vast quantities for removable storage devices: CompactFlash cards, USB "pen disks," Secure Digital (SD) cards, MP3 players, and so on. Although it is possible, with a minimal amount of external logic, to graft NAND Flash onto a normal Flash/ROM/SRAM controller, there are a couple of reasons why you can't simply boot directly out of the NAND Flash. The first is that NAND is not guaranteed error-free; it's the host's responsibility to maintain ECC and bad sector mapping information. The second reason is that NAND Flash is addressed serially; you send the chip a block number, then read the block out into RAM. Hence, you need a little boot firmware on a normal random-access PROM to do the physical-level management of the NAND.

Note that some microcontrollers provide hardware NAND controllers that obviate the need for the little boot PROM I discussed previously. The disadvantage of relying entirely on that sort of hardware is that you lose the failsafe system-recovery features that can easily be implemented in the boot PROM. However, if you're working against space or cost constraints, and your micro has the NAND control hardware, you may want to avail yourself of it. SoCs sold for cellphone applications use this sort of technology.

Hopefully you've found this to be illustrative of some of the initial things you'll need to consider when embedding Linux. One of the great points in favor of this operating system is the excellent community support; whether you're an individual trying to bring the OS up on a new platform (and do some learning in the process) or simply writing some application software, there is very good support available on Usenet and message boards. There are so many people using Linux in so many different ways that you're almost certain to find someone who can help you answer your questions.

4.6   eCos

eCos is an embedded operating system designed for smaller applications than Linux. (For some reason, people seem to have a lot of mental trouble separating these two operating systems.) If you're working on a small hardware platform, or you don't wish to get into the GPL-ness of Linux, eCos might be the right open-source choice for you. Much like Linux, eCos is supported by the gcc toolchain, so it doesn't cost anything to start experimenting with this operating system. Although it doesn't support anything approaching the massive range of hardware covered by Linux, eCos has been ported to a respectable number of (larger) cores, including x86, ARM, PowerPC and several others.

The eCos homepage is *<http://ecos.sourceware.org/>*. Probably the first thing you'll want to read there is the license—eCos is royalty-free, and covered by a license that does not require that you disclose your modifications to the general public. As you will see from the list of supported hardware, eCos has been ported to about a hundred different hardware platforms, with all the major CPU cores you're likely to use covered—there's a good chance that you won't need to do much porting work to get it operational on your own hardware.

In a nutshell, eCos is a POSIX-compliant multithreaded operating system with strong real-time capabilities, networking support and various other useful features such as USB (slave), serial and Ethernet support. Some ports also include the code to initialize a display device (for example, the port to the evaluation board for the Cirrus Logic Maverick EP7312 evaluation board), although the core OS does not have any GUI support built into it.

eCos is not Linux, though the two are frequently confused—since Red Hat supplies flavors of both, people somehow assume that eCos is derived from Linux. This is explicitly not the case; the eCos maintainers are careful to keep Linux code out of the source tree, to avoid having to put eCos under the GPL. The main advantage of eCos over Linux (and many other operating systems) is greatly superior real-time characteristics.[11] eCos is also configurable to a very fine level of granularity; merely by selecting a few configuration options at build time, you can prune the operating system's driver set down to the bare

---

[11]   When compared with standard Linux. Linux with real-time extensions is, of course, much better than regular Linux at meeting real-time requirements.

minimum required for your application. While the Linux kernel is, of course, configurable—and you can pull some functionality out of the base kernel image and put it in dynamically loaded modules if necessary—the baseline memory requirement for Linux is still much higher than for eCos.

Note also that eCos is a multithreaded operating system, not a multitasking operating system. Your application is not "loaded" in the same way that Linux loads external programs off a hard disk or out of Flash. In eCos, your application is directly linked in with the operating system; the operating system is actually a library and a body of startup code that is statically linked with and hence inextricably welded to your code. The possible downside to this is that, unlike in Linux, in eCos you can't directly spawn an external program to do some function for you, since the operating system has no concept of "loading" a program (and even if it did, there is no defined API for such an externally loaded program to access operating system services). If you want to include the functionality of some external utility, you will therefore need to integrate the sourcecode of that utility directly into your own code and statically link it into your operating system image.

From a commercial standpoint, another advantage of eCos over Linux is the license agreement. Some investors in particular might be leery of using Linux in a commercial application because of the GPL (though there are certainly many precedents for it, many of them from very large names indeed; there's no accounting for irrational fears). With eCos, you get all the license advantages of a payware commercial embedded operating system without the price tag—and you get a powerful, multiplatform, open-source operating system to boot.

There are a few reasons why I bring up eCos in this book. The first reason is that it is really an excellent teaching platform for migrating consumer-type programming skills into an embedded environment. I can think of almost no exercise better suited to the development of these skills than porting eCos to a new hardware platform. Second, eCos is more or less a superset of RedBoot, RedHat's highly flexible bootloader. RedBoot is a *really* useful product to have on your system, because it lets you load software from a variety of different interfaces and write it to Flash or other internal storage. It's worth learning how to develop for eCos (which uses the same hardware abstraction layer) merely in order to be able to get RedBoot running on your hardware—it's one of the two most powerful open-source bootloaders available (the other being U-Boot, which I mentioned in the section on PowerPC). Although RedBoot is often mentioned in conjunction with eCos, it can be used to load any embedded operating system you care to name—Linux, eCos, or something proprietary you write yourself.

Finally, eCos is a powerful and viable operating system in its own right—perfectly acceptable for fielded consumer products, as you'll see from Red Hat's list of design-in success stories. It has a very swift boot time (especially compared to Linux), relatively small memory requirements, and overall it makes an excellent choice as the operating system for a 32-bit project, particularly a network-connected project.

The only book on eCos that I'm aware of is Anthony Massa's *Embedded Software Development with eCos* (Prentice Hall, November 2002, ISBN 0-1303-5473-2). This book is an excellent reference, and I do recommend it, but it's getting rather old now—you will definitely want to seek further support in the eCos discussion forums.

If you're an experienced C/C++ programmer and you want to build a really impressive 32-bit embedded application, then I suggest that eCos will let you reach this goal quickly and easily. In almost every aspect (except perhaps GUI development), you will find eCos easier to port than Linux, and also considerably more scalable to small devices. If you do need that GUI support, then you should look at MiniGUI, *<http://www.minigui.org/>*—this product has explicit support for eCos. It's not the only option for implementing a GUI on eCos, but at present it seems to be one of the best-supported paths.

4.7 What Programming Languages Should I Learn for Large Embedded Systems?

When the phrase "large embedded system" is employed according to the definition I've had in mind throughout this chapter, all large embedded systems will use an operating system of one kind or another. The custom software bundle loaded on such a system is not usually monolithic, either—there will likely be an application layer that is quite distantly abstracted from the hardware, some device drivers for your custom hardware, and probably some software in between that isn't exactly application code but doesn't live in the driver layer either. (If you've read my previous book, *Open-Source Robotics and Process Control Cookbook*, you would have seen me talking about lircd, the daemon that decodes incoming infrared remote control signals from the infrared hardware driver on Linux systems, and translates the codes into developer-defined strings. Lircd is an example of the sort of "in the middle" software I'm talking about here. If you're unfamiliar with how it works, you can read all about it—and download it—at *<http://www.lirc.org/>*.)

As a result of all these layers, it is likely that any such system will contain significant amounts of code written in several different languages. On the other hand, however, the sheer size of the project is such that the software team will often be large and not necessarily cross-functional. Consequently, the exact skill set you'll need depends on where in the abstraction hierarchy you wish to work. Very generally: if you want to work in the lowest level of device drivers and boot-loaders, you'll need good assembly language and C skills. You'll also need to know how to operate the hardware debugging tools; JTAG adapters and so forth.

If you're working in the middle layers, the skill set is potentially quite fuzzy. Some of these applications are written in relatively exotic languages like perl and Tcl; many of them are simple C programs. In this part of the hierarchy, your most important skill will be an intimate knowledge of how the operating system's skeleton is constructed and what services the applications upstream of you require. You might also need to be able to drive an oscilloscope and/or some kind of protocol analysis tools in order to debug what's going on with your code, since you might be poking some device drivers and/or tinkering with arcane aspects of the operating system's state.

At the upper level of the hierarchy, the required skills are dictated by the nature of the end product. You might be working in Adobe Flash, or Java, or C++—it depends entirely on the product. Skills you've learned programming consumer operating systems in the application layer will be very useful here. You probably won't be debugging your program using hardware tools as such; possibly you'll run a debugger stub on the target and talk to it over a serial or Ethernet link to see what it's doing. It is quite unlikely that you'll need to use any hardware debugging skills.

By the way, you shouldn't take all this talk of large teams and programmers working inside insulated cocoons as being unalterable gospel by any means. I can tell you from personal experience that some high-end 32-bit systems for consumer products are developed entirely by a single engineer. The reason I'm focusing on the big-team picture here is because of the target demographic I announced at the start of this chapter—people who are most experienced in the application layer. I'm trying to point out how these enormously complex systems can be made accessible to the one-person team with Java experience but not much understanding of real-time systems. If you have—or want to gain—experience with the lower levels of the software hierarchy, there's no reason why you can't build your own completely custom 32-bit system, and some of the resources I've mentioned in this chapter will be useful for you in your pursuit of that goal. However, in addition to all the high-level stuff I discuss in this chapter, you'll need to have a good understanding of the lower-level details, as well as the sorts of selection criteria that I mentioned in the previous chapter, where I was talking about much more "down-to-the-metal" systems.

4.8 A Final Word on Part Selection

In this section, I'll try to give you a little taste of the joys of selecting a high-end microcontroller for an embedded application. The challenges and choices here are often quite different from the process you'll follow when selecting an 8-bit microcontroller. The stakes are also higher, because the design effort invested into a 32-bit circuit is likely to be significantly larger than the design effort in most 8-bit designs.

The issue that makes 32-bit device selection potentially so difficult is that if you're considering these high-end parts, you quite likely have a fairly complex application to implement. You're then faced with the task of deciding from a datasheet whether or not a particular part can do whatever you need, and this is often very difficult, particularly because at the pre-prototype stage, everything is up in the air and the requirements themselves are subject to change. Often, what you'll be given is a be-all-end-all wishlist that can't possibly be achieved in a single design while remaining cost-effective. Hence, you'll generally be expected to present a few options to the marketing staff, who will then pick the one that they think will sell the best.

By the way, there's just one not-quite-so-rare exception to this situation: in a few lucky cases, your reason for choosing a 32-bit part is simply for the extra address space, if your application requires lots of directly addressable memory for some reason. If your device falls into this category, thank your lucky stars!

I'll illustrate the point of this section with a practical example from my own career. Some years ago, I was asked to design a new, greatly enhanced version of a multimedia appliance. The bluesky wishlist was amazingly long, and included the following items:

- High-resolution still image display with fast decoding time and the ability to apply special effects and animation as the new image was being placed onscreen.

- DVD playback.

- Support for MPEG-1, MPEG-2, MPEG-4 and QuickTime movie files, as well as MP3 audio.

- Data CD and video DVD burning support.

- Internal hard disk.

- Wired and wireless Ethernet.

- The ability to receive pictures over Bluetooth.

- Infrared remote control.

- The ability to play Macromedia Flash movies.

- Fully scriptable operation.

- Support for all the popular digital camera Flash media cards.

- The ideal circuit should be able to drive all sizes of LCD panel from about 5.7" diagonal, QVGA resolution (320x240) to 23", SXGA resolution (1280x1024) at 24bpp.

- Ideally it should be possible for the small-screen versions to be battery-powered.

As you can doubtless see, there are numerous demands here, some of them conflicting. Thus, I started the quest by looking at the most obvious user-visible feature, viz. the video hardware. As it turned out, this was a good choice for first examination, because this feature channels the entire design of a product of this type. The difficulties were manifold: To drive small displays (up to about VGA resolution, 640x480), the most cost-effective solution is to use a highly integrated System-on-Chip (SoC) microcontroller with on-chip LCD controller. Several vendors offered us parts that would work reasonably well for low resolutions. However, these devices almost all employ a unified memory architecture; RAM bandwidth is shared between code execution and DMA peripherals such as the LCD controller. As you increase resolution and color depth of the video output, you drastically reduce the bandwidth available for executing code. This meant that for anything larger than about QVGA size, the CPU would be so starved for RAM bandwidth that it wouldn't be able to handle the video decoding tasks (even a more modest subset, like MPEG-1 at VideoCD resolution, let alone trying to do a full software decode of DVD-resolution MPEG-2). Furthermore, there wasn't a single SoC-type device that would handle the higher resolutions desired for the large-screen models.

In order to support those higher resolutions, therefore, it was necessary to use an external video chip. Unfortunately, all of the suitable parts are designed for use in laptop computers. Not only does this mean that it's very difficult to get technical documentation on these parts, but they're also virtually impossible to buy in small quantities (where "small" is defined as "less than a few hundred thousand per year"). Perhaps worst of all, they all require PCI or AGP bus support; none of them are designed to be tied easily to microcontrollers. Relatively few micros support PCI natively, since it's not required for the vast majority of embedded devices. You can buy various PCI-to-"other" bus interface parts, and you can also roll your own out of an FPGA (some vendors provide FPGAs with PCI support explicitly built in on one end, specifically for this application), but it's a real bear to implement this.

Additionally, all of the laptop video chipsets of the present day are shipped in BGA packages, which we couldn't prototype (a complete set of assembly and inspection equipment is rather expensive)—so the prototype manufacture would have to be outsourced. In order to support the large panels on the same hardware as small screens, we were going to have to commit to a high-end, power-hungry microcontroller (implying short battery life for the models with battery power), with on-chip PCI interface. We were also going to have to design in a hard-to-source laptop video chip with a very short lifespan driven by the consumer PC market.

We eventually decided to use a single-board computer with most of the desired functionality on-board rather than trying to design our own circuit, mainly due to the prototyping and sourcing difficulties. However, when trying to select an appropriate SBC, we ran into a benchmarking problem with respect to the full-motion video. We had some approximate metrics gathered on PCs, but benchmarking video performance is rather unique; you really need to evaluate, simultaneously, the CPU, memory subsystem, disk or network interface (depending on how the compressed video stream is going to reach you), the acceleration support provided by the video chip, and the device driver's ability to exploit those features (in other words, just how well does your operating system support the hardware you intend to use?). We were going to run Linux on the device; some SBCs had very powerful CPUs, but their video chipsets weren't well-supported by Linux, others had great Linux support all around but their CPUs were a bit too slow for the required still-image animation tasks. We never found a board that would properly support the large 17" panels directly (these panels use multichannel, multipixel-per-clock LVDS interfaces, which aren't well-supported on laptop chipsets; they're intended for desktop monitor applications).

As a matter of interest, we wound up compromising on a lot of features—we temporarily shelved the idea of the really small models and decided to focus on larger devices using an analog video path to avoid the utter incompatibility of the laptop video chips with the larger panels. The product was, ultimately, commercially successful. However, the initial idea we had had—namely, that we would take a StrongARM microcontroller and meld on an ASIC or DSP to handle the video decompression—was totally discarded. In fact, we built something that very closely resembled a panel PC, with an x86-compatible single-board computer inside it. Hopefully this illustrates to you to at least a small degree how a high-end requirement can lead to "you can't get there from here" design problems.

5

Working for Yourself as an Embedded Engineer

5.1   Is Self-Employment for You? Risks and Benefits

More or less once every quarter, I get together with a group of engineers and miscellaneous IT workers (website designers, database programmers, fuzzily named "technology consultants," and so on). We have dinner and swap stories of managers, contracts and customers. Once in a while, a newbie visits the group to network and hear tales of the real world while they decide whether to start moonlighting. Most of these newbies subsequently decide to stick with their day jobs.

Working as a full-time consulting embedded engineer is a difficult job. If you've only worked in a company environment before—even a relatively small company—you probably have no idea just how much infrastructure helps you through every hour of the day. If you're tired of process and procedure, if your day job leaves you exhausted and you're and looking for a life of freedom through self-employment, I'm afraid I have some very harsh news for you. *It's substantially more work and (at least intermittently) more stressful to be successfully self-employed than to hold down a day job. Furthermore, if you don't organize yourself and work to an established procedure, you'll lose customers because of silly omissions or mistakes.*

The key difference between working for "the man" versus working for yourself is that self-employment lets you make your own choices about which opportunities to pursue. Here are some good reasons to start working for yourself.

- You are currently earning *at least* twice your day-job hourly rate doing contract work as a moonlighter, and you want to focus all of your attention on where the money is. (More on this important topic later.)

- There are occupational health concerns (for example, too much overtime work) or insurmountable personal conflicts at your day job. Working for yourself is far from stress-free, believe me, but at least you'll find it hard to argue with the company.

- Your day-job employer is in poor financial health, is moving out of the marketplace for the products you engineer, or is progressively outsourcing more and more of your department or division.

- You're bored with working on maintenance of legacy code and circuits and you want to work in a job where you'll be exposed to more diverse projects.

- You like being involved in the entire product development cycle from concept to implementation to marketing.

Here are some fundamentally bad reasons to start working for yourself (and by "bad," I mean in the sense that you're probably not going to achieve your stated goals):

- You dislike a structured approach to product development.

- You like building prototypes for quick demonstrations but hate the I-dotting and T-crossing details of designing something that's reliable and manufacturable. There's a big difference between hand-building a single prototype and designing the same type of product to be manufactured by automated machinery, or by semiskilled labor.

- You want to work shorter hours.

- You hate paperwork.

As with any other major life decision, becoming a full-time consultant is a complex choice. On the plus side, you will have about as much control over your professional destiny as it is possible to have. You'll be able to set your own policies, enforce your own design style, and select precisely which projects you want to pick up and which you want to leave on the curb. You will also have unlimited access to the (after-tax) profits of your labors; there will be no management layers or investors getting a slice of the pie. Being a consultant can feed some very interesting projects your way, full of enjoyable technical challenges and—ultimately—often considerable profits. It can be an end in itself (many engineers "retire" into consultancy as supplemental income and to keep their brains in top condition), or it can simply be a way of looking around you and keeping your skills honed and the bills paid while you decide what your next full-time job should be. Consulting is a great way to network, by the way. People who are looking today for a consultant to work on their first electronic project will be looking tomorrow for someone to work full-time on their range of electronic products.

The downside to being a one-person show is that almost every decision you make when you're working for yourself can mean the difference between life and death. This is frightening, and with good reason. If you're working for yourself, you won't have anybody with the same personal investment as yourself to help you with the myriad of decisions and tasks that are needed to make a business of any size function as a going concern. If you're working as a sole proprietorship (as opposed to an incorporated company), you might also have to worry about personal liability if one of your projects doesn't turn out favorably. You'll have to balance the cost of advertising, installing a fax line, and/or buying a piece of equipment against the profits you expect to get. It is inevitable that you will make some mistakes here, and those mistakes will cost you some money.

If all of this hasn't scared you off yet, congratulations, and please continue reading.

5.2 From Moonlighting to Full-Time Consultant Status—Bookkeeping, Taxes and Workload

Keeping track of your finances, ensuring that you're actually turning a profit, and paying your income tax in a legally acceptable manner can be one of the most difficult and frustrating tasks of being a freelancer. What I'm about to say here is structured around tax terminology and laws in effect in the United States, but most developed countries have very similar laws. Please be aware of the obvious disclaimer: I am not an accountant, and this text does not constitute formal tax advice in any way. Furthermore, tax law is a very complicated, specialized field and it's constantly changing. You should consult a CPA or other tax professional for detailed recommendations about your specific situation; these people are paid to have an up-to-date understanding of tax rules. Shelling out for a few hours of your CPA's time four or five times a year (once per quarter, and perhaps an additional visit during tax season) is a much better value than spending considerably more of your own time trying to learn about the topic—not to mention the fact that if you make a mistake, you can dig yourself into a deep, expensive hole with a huge tax bill and penalties on top of it.

The statements I'm making here are redacted from several references, primarily IRS publications for the 2005 financial year. Before we go any further, you'll find the following glossary helpful (particularly if you've never filed a tax return in the United States).

- *1040* – This is the form (actually, a family of forms) on which you submit your annual tax return.

- *1099* – This is a form that is mailed to you at the end of every tax year by each person that has paid you more than $600. Again, there are in fact several different flavors of Form 1099, including 1099-DIV for dividend income, 1099-MISC for miscellaneous income, and so on. Clients for freelance jobs will generally send you 1099-MISCs, if they send you anything at all. When a 1099 is generated by a person who paid you nonsalary income during the year, one copy is sent to you and another goes to the IRS. You don't need to include these forms with your tax return, since the IRS already has the information; you merely need to declare the totals in the correct lines on your tax return.

- *AGI* – Adjusted Gross Income. This is essentially your gross income less allowable deductions; i.e., it's the actual "income" number on which your tax bill is calculated.

- *FICA* – Acronym for Federal Insurance Contribution Act. This term is most commonly used when referring to withholding for Medicare and Social Security.

- *IRS* – Internal Revenue Service, the body responsible for collecting federal taxes.

- *W-2* – This is the standard form issued annually by regular ("day job") employers stating, among other things, how much was paid to you and how much Medicare, federal, state, and local taxes were withheld from your salary. Forms W-2 must be included when you submit your tax return.

I'll assume for the moment that you're going to follow the usual route of 1099 income—either doing some moonlighting contract jobs or full-time freelancing (as distinct from setting up your own corporation and having that corporation pay you a salary). Freelance work is almost always offered on a 1099 basis because it's less expensive for the person who hires you; there is no employer–employee relationship. Employers who contract with too many W-2 employees—even if they're all part-time semi-freelancers—start to appear on the radar of various industrial relations legislation that can be quite irksome to deal with.

Important warning: There are an astonishingly large number of resources available—especially online—that discuss ways to structure a home business so that it runs permanently at a loss that can be offset against your normal W-2 income. The underlying plan here is, in a nutshell, to make your hobby a tax-deductible expense.

Unfortunately for you, these methods are questionably legal, since at best they severely stretch the definition of what constitutes a "business" expense. They are definitely not sustainable in the long term—the IRS has various rules of thumb to classify home businesses as nondeductible hobby activities, but the primary criterion is summarized by the questions "Does this activity generate a profit? If not, is it being carried out in a manner that implies the expectation of future profits?" If the answer to the second question is "no," your deductions will be disallowed and you'll be on the hook for a hefty tax bill, with penalties. Since a consultant engineer's primary raw material is work hours (rather than paid-for physical raw materials), it can be amazingly difficult to make a freelance engineering business run at a net loss anyway, so you might not even be able to make it work on paper, let alone pass IRS scrutiny.

If you really want your pet project to be tax deductible, I've got a suggestion that sounds flippant but really isn't: Find someone who will pay you to do this pet project. One excellent way to do this is to write about it and get the articles published—that way, your raw materials become necessary research expenses.

The long and short of all this is that you should assume that your freelance business is going to turn a profit, and work on that basis from the beginning.

Of course, if you make a loss, you're entitled to all the tax "benefits" of such a loss—but be prepared to answer questions if your business doesn't start to run in the black within a couple of years. Don't throw out a single scrap of paper!

With these factors in mind, we immediately run into something of a financial difficulty: the IRS doesn't want to wait until the end of the year to get their cut of your earnings. They want to be sure that you've paid at least 90% of your tax bill by the end of the year. For "day job" (W-2) income, you don't have to worry about this—your employer withholds the correct amount from your paycheck and distributes it to the IRS and your state government's tax department (oh yes—you have to pay state income tax too, I'm afraid). Self-employment income, however, is not subject to withholding; the person who hires you just writes you a check for the amount you agreed to charge for the job, and paying the tax on that is your responsibility.

There's a second, related issue to worry about with taxes on freelance income, and that's FICA. The federal government wants 12.4% of your AGI, up to a certain limit, for Social Security. They want another 2.9% of your AGI for Medicare (and there's no cap on the Medicare tax). On W-2 income, your employer pays half of this amount; for self-employment income, you have to pay all of it. That's one reason why freelance income needs to be relatively large if it's intended to replace regular day-job earnings—you don't get to keep as much of it. Note, however, that you do get a slight break—if you have to pay the full FICA amount (self-employment tax), you can claim the half that would normally have been paid by your employer as a business expense.

If you're earning any significant amount from activities that don't withhold income tax, you need to pay quarterly estimated tax using form 1040-ES. The worksheet on this form is, to my mind, obscenely complex, because it basically requires you to guesstimate most of what will be on your tax return at the end of the year just now beginning, without the benefit of any of the paperwork that you'll receive during the year. Note that there is an exception to the requirement to file income taxes: if you expect that your withholding for this year is going to be at least as much as you paid, total, in tax for the previous year, you won't be penalized for failing to pay estimated tax on time.

There are numerous strategies you can use to manage the problem of estimated taxes. The method that's best for you will depend, among other things, on how much you expect to earn from freelancing, how accurate those predictions are, whether or not you have a day job as well, and how finely tuned (or strained) your week-to-week finances are.

It's easiest to manage withholding versus 1099 income if you have a W-2 job that earns you enough to live on with a reasonable margin. If this is the case, you can simply get your employer to withhold more from your paycheck so that at the end of the year you're approximately up to date with your tax payments. You can roughly estimate how much has to be withheld by using the following method:

- Estimate the net amount you expect to earn from 1099 activities over the year (estimated income minus estimated 100%-deductible items such as parts for prototypes, telephone expenses, and so forth). This can be tricky unless you have your entire year's work planned out in advance, which is rare.

- Work out your marginal tax rate. The IRS publishes marginal tax tables at the start of the year; you can read the table for 2006 at *<http://www.irs.gov/ formspubs/article/0,,id=150856,00.html>* but in essence there are six brackets; 0%, 15%, 25%, 28%, 33% and 35%. Which bracket your income belongs to depends on how much you earn and your filing status (single, married filing jointly, married filing separately, or head of household); the details are provided by the IRS. Work out how much federal tax (we'll call it $T) you owe on the extra income by multiplying the percentage rate by your estimate of additional income.

- Work out how much FICA you owe on the additional income, by multiplying the estimated dollar amount of 1099 income by 15.3%. We'll call this amount $F.

- The total additional tax you owe for the year is going to be very approximately $T + $F,[1] so you can work out how much additional tax your employer should withhold per paycheck by dividing (T + F)/(number of pay periods in a year).

---

[1] This is a very rough estimate, of course, and doesn't take into account all sorts of potential deductions, but it's close enough for our purpose. More important, it's a very conservative number that's likely to result in overpaying the IRS. The goal here is to avoid a tax bill at the end of the year.

Example: Suppose you're single for tax purposes, and you earn $50,000 per year in your day job, which pays you on a biweekly basis. Further suppose that you believe you're going to get contracts worth $10,000 and you're not going to spend anything on supplies or other deductible items in order to make that money. According to the 2006 tables, you're in the 28% marginal tax bracket, so you are going to owe an additional $2,800 in federal tax on that bonus income. You will also owe $1,530 in FICA costs, for a grand total of $4,330. Since there are 26 pay periods in the year, you should ask your employer to withhold approximately an additional $167 per paycheck.

The pain of paying self-employment tax is by this means reduced to a chronic ache spread evenly throughout the year rather than an acute attack in April (assuming you predict your annual income with reasonable accuracy, of course). This method doesn't work very well, however, if your expenses are already very close to your take-home W-2 pay, because if there's a fluctuation in the stream of incoming 1099 cash, you might not be able to meet daily expenses. In such cases, you're better off going the traditional route of filing estimated taxes on a quarterly basis. Again, the simplest way of doing this is to sidestep the 1040-ES worksheet and simply to take your gross "extra" income, subtract the deductible expenses from the quarter, and calculate the tax and FICA you owe on the remainder. You'll be overpaying up-front, but if you compare the lost potential interest on that money with the cost of calculating the "right" amount of estimated tax, you might still be ahead of the game. If your tax situation is even slightly more complex than this, however—for example, if you have incorporated your business—you really need professional assistance.

If, in addition to freelancing, you're working at a day job with a 401(k) plan (variously referred to as a superannuation account or retirement fund in countries other than the United States), one of the most important things you can do with your additional moonlighting revenue is to maximize your pre-tax 401(k) contribution by increasing the contribution percentage taken out of your paycheck and making up the shortfall in your day-to-day living expenses using your 1099 income. You get two benefits from this. First, if your employer has a matching scheme, you literally get extra free money. Second, you get to sock this extra income tax-free away into a location where it's going to grow interest for you on money that you would otherwise be paying out in up-front taxes.

Note that for 2006, the maximum pre-tax 401(k) contribution is $15,000. If you're over 50, you can make an extra "catch-up" contribution of up to $5,000. If you're getting close to those limits, you'll need to look for other investment vehicles to store the extra money; it's a nice problem to have.

In order for any of these suggestions to work, you need to keep very good records of how much money you bring in and how much you spend on business-related expenses (and also what those expenses are). The reason it's particularly important to keep track of your income is because you will VERY frequently not receive all your 1099-MISCs in time to use them to calculate your taxes, especially from small customers—or at least, that has been my experience.

My surmise is that this occurs because many small businesses leave tax preparation right down to the wire. In early April, they visit their CPA and show the CPA records saying they paid you so many thousand dollars during the year (which, of course, they want to claim as a deductible expense). The CPA generates 1099-MISC forms while preparing your customer's tax return, and then hands the whole bundle to your customer. The tax return is mailed in urgently; the 1099-MISCs are mailed out on a more leisurely schedule, if at all. Technically, all that paperwork is supposed to be in the mail by January 31, but this rule seems to be more or less routinely ignored.

Although the previous paragraph is admittedly conjecture on my part, the fact remains that you need to keep your own records of income received and not rely on getting everything in the mail before mid-April. This brings me neatly to my next major point: You need to have good records of *everything*. There are a myriad excellent reasons why it might be very important to find out exactly what you charged for a job three years ago, or to locate the first version of a schematic for a project you did six months ago.

You can only keep all this information at your fingertips if you put in the time to create and actively use a functional filing system. This can be primarily electronic, primarily paper, or a hybrid of the two. Since you'll inevitably be receiving paperwork from customers, suppliers and other people, the hybrid system is probably easiest to manage. Some people really like to keep everything in electronic format, so they scan their phone bills and other correspondence, shred the paper originals or stuff them in boxes into storage, and convert the scanned copies into Adobe PDFs or some other convenient format. Some people, on the other hand, are paper-lovers; if they receive an electronic invoice, they'll print it and file the printed copy.

I personally favor the hybrid approach. All my correspondence is generated electronically, as you might expect, so it's archived electronically. Check stubs, incoming letters and bills are archived on paper, as are signed contracts and various important drawings, schematics and so forth (although I do, of course, keep the original electronic CAD files as softcopy). I use the free GnuCash software *<http://www.gnucash.org/>* to keep track of my finances (it's a simple double-entry bookkeeping system). However, at the end of the day it doesn't matter a whole lot if you go electronic, paper or some route in between the two. What's important is that you put everything into your filing system, and you do so in a coherent way so that you can find filed items again when you need them.

In fact, I advise you think about how your filing system works and write down a document describing it. You'll be able to use this document to train a secretary, if you get to the point of needing one—and formalizing the manner in which your business operates will greatly help you to follow your own procedures consistently. It's also the underlying core of—dare I say it—ISO9000 certification, if you ever need to travel that route.

5.3   Ways to Find and Keep Customers

One of the reasons it's easier to start your working life with a day job and transition into consultancy (assuming that's your ultimate goal, of course) is that it's very difficult to build instantaneously a sufficient base of clients to sustain a business. For this reason, if you're contemplating a life of full-time embedded consulting work, I advise you to give serious consideration to holding at least a part-time job to help you through the lean times, especially while you're getting started. Additionally, I would advise you not to consider consulting work, even part-time, until you have had *at least* two or three years' experience working in the industry. Preferably you should get this experience at the engineering level, but it's an acceptable second best to have experience from the manufacturing side, since you'll be exposed to a lot of the same learning curve.

You can feel free to ignore that last piece of advice, of course—some people do get lucky, and jump straight into successful consulting positions right out of school. However, even after you have fully imbibed and taken to heart every word in the invaluable book you are now reading, you will not be fully cognizant of just how frequently, and how spectacularly, things can go wrong in one-man embedded projects.[2] School can give you the basic rules for designing circuits and writing code, but (in my experience of working with fresh graduates) it does not teach realistic project management skills. Project management includes all sorts of problems you won't encounter at school; dealing with technical support for the chips you've designed in, doing PCB layouts (many EE grads have never laid out a board; as a consultant, you'll probably want to do this yourself for profitability reasons), making sure that all the parts you've specified in are available and poring through catalogs to get the exact part numbers and pricing, and a myriad other tasks.

Nor does a course of schoolwork leave the average student with enough completed projects under their belt to have an intuitive feeling for where the project should be overengineered for future expansion, and where it can be designed down to the last millivolt.[3] As a result, when you're new to the field, you will find yourself chronically underestimating the time and cost of the work you're asked to perform. Customers want results—delivered projects—not merely activity. If your food budget relies on the income from this work, you're potentially in a bad situation if you're unable to make accurate estimates of how many projects you can get completed in a given time period. You can partially compensate for this by making sure you charge by the hour (see Section 5.5 for more details on this), but you won't be making good long-term friends out of your customers if 60% of your billed hours are spent correcting problems you created by under-specifying the original hardware.

---

[2]   I don't just mean in the sense of encountering bugs that are difficult to fix. The kinds of problems I'm talking about are issues like your client deciding that some new feature is a must-have, right after you spent a week of effort laying out a PCB around a microcontroller that doesn't have enough RAM to implement the feature, or finding out immediately before production is due to start that your PCB layout is just a shade too noisy to pass CE emissions requirements.

[3]   As a prime example of this, all other things being equal, it is preferable to choose a microcontroller that is available in pin-compatible variants with differing amounts of ROM and RAM. This way, when you find yourself five bytes short, you can simply upgrade to the next micro without needing to create a new PCB.

Given that you have all the requisite experience, how can you find people to buy your services? The usual way of selling anything is through advertising, but you won't sell much in the way of embedded engineering services by buying a full-page advertisement in the New York Times. There are a few obvious methods of advertising this type of service, and I suggest you try a combination of them.

- *Build a website detailing your capabilities and keep it up to date.* In this day and age, it costs practically nothing to host such a site (and the cost is tax deductible, anyway). While simply having a website won't spontaneously generate business, it's important to have a place where you can direct people to read more about what you do. Having a reasonably attractive website with all the pertinent content customers are going to want to see also helps to create a professional appearance for your operation. On the subject of attractiveness vis-á-vis websites by the way, my overarching advice is: Don't go crazy. I suggest using plain HTML with no frames, no animated graphics, and of course you should **never** bury any of your content inside Flash movies or Java applets. [4] I'd also suggest eschewing embedded scripting content, particularly in the navigation aspect of the site. Having a few rollover graphics that change color via Javascript is not a mortal sin, but forcing the user to use Javascript-powered drop-down menus to get off the homepage is an extremely bad idea. You want your website to be quickly navigable by visitors; you want it to look good without requiring any plugins or magic browser features; and perhaps most important, you want it to be easily indexable by search engines. Make sure you include pictures and descriptions of projects you've completed. (Personal and school projects are fine, though once you have a significant portfolio of actual customer work, you will probably want to deemphasize the older material.) If the numbers won't embarrass you, it's helpful to show people how long each project took to complete.

---

[4] This is partly the engineer in me speaking. These animated graphics tools are designed for style, not substance. Hiding content inside these binary files is a very silly choice, if only because it will exasperate most of the engineers who will look at it. You're not selling lipstick to teenage girls, so you don't need singing cartoon ferrets on your website. You're selling professional services to people who have limited time to read, understand, and evaluate their options. Don't make it difficult or annoying for them to read about what you have to offer.

- *Publish as much as possible.* There are a large number of specialized engineering publications[5] that are distributed free (in both paper and electronic form) to practically anyone who fills out a survey and is willing to endure endless junk email. At the risk of sounding unkind, the publishers of these materials need a modicum of actual content to wrap around their advertising. If you have the ability to write an interesting article about something relevant to your industry segment (typical article lengths are in the region of 2500 words), being published in these magazines is excellent free exposure. You'll also usually be paid a small amount for the article; not typically enough to match normal consultancy rates for the same number of hours, but the bulk of the value in the transaction is the free advertising you get, and the actual cash is just a pleasant bonus.

- *Release free project information.* Wherever possible, release code and schematics for your personal projects (put it up on your website). You're improving your visibility in all sorts of ways here—as people link to your projects, your search engine ranking increases, for instance. You're also increasing the probability that someone looking for help with a particular sort of widget is going to find their way to you.

- *Solicit local small businesses that might benefit from your services.* In particular, if you have a personal or school project that will demonstrate experience with a particular industry, it's a great idea to include details of that project with your letter.

- *Attend conferences for networking purposes.* Because of the broad scope of embedded engineering, there are literally hundreds of conferences where you might meet people that can use your talents. You'll get advance notice of many of these conferences (perhaps obliquely) if you sign up for newsletters from the various semiconductor vendors that attend these events. Some others, like the annual Toy Fair in Manhattan, might require a bit more research on your part.

- *Create an unusual electronic promotional item and send it out.* This approach should only be used on very good prospects, because it's usually quite

---

[5] Advertising circulars with a little content added in order to get people to open them, mainly. A kinder phrase for them is "industry publications." I'm not the only one to hold this cynical attitude; industry news in the Internet age arrives faster than the leadtimes for these print publications.

expensive. A typical example of the sort of thing I mean would be to make an electronic business card with, say, LEDs and a small microcontroller on it; maybe a piezo speaker as well.

So much for acquiring new customers. Retaining existing customers is an interesting problem for the consulting embedded engineer. Quite often, the reason a company goes hunting for a consultant in this field is to make a one-off project that is outside their normal field of operations. For example, due to my former involvement in the toy industry and a magazine article I wrote on the topic several years ago, I get a lot of phone calls from wannabe inventors needing expert help to make a prototype for submission to a big company. I also get a fair number of calls from marketing people who are building flashing or speaking in-store displays, and have no idea where to start with the electronics of such a device. Since I don't work in the industry any more, I refer all those sorts of calls to colleagues, but I suspect that relatively few of them lead to much repeat business; they're one-time things.

Unfortunately, there's not a lot of advice I can offer that will help you persuade people to start new projects. Your best course of action is very simple: complete your work in a timely and professional manner, following the suggestions in Section 5.6 along the way. During the course of the project, offer intelligent commentary and advice for improving the end-user experience for whatever device it is you're making. This sort of useful engineering advice is precisely the sort of value-added bonus that will keep your name at the top of your customer's Rolodex.

It certainly doesn't hurt to follow up with your customer a few months after delivering the completed product. By this time, the device will have been in production for a while, and if the customer hasn't called you already, you can be fairly sure there are no show-stopping problems. A couple of months' real field experience will quite likely, however, have exposed some rough edges; things that can be improved, or accessories that can be created, and these are additional business opportunities for you.

Some people even advocate calling through your client database whenever things are a bit quiet, to see if there's any business to be captured. I'm slightly uncomfortable with this approach; in my view, it's bordering on the unprofessional.

However, if you do decide to go down this route, try to keep the conversation from becoming a blatant sales call. Ask your customer how their older projects are behaving in the field, and ask them if they have anything new coming down the line by all means. You'll quite likely have to spend some time discussing the pros and cons of various ideas they are mulling over at the moment. Anything much beyond this is begging. The bottom line is that your customer either does have some new project ideas, or they don't. If they do have something in mind, then simply by calling and talking for a while, you've reminded them that you exist and you've given a strong hint that you're available to work. If, at that point, they don't want to hire you for the job, then their quite likely unhappy with your past performance for some reason. Trying to repair that situation can be difficult, especially if your customer doesn't want to talk to you about it.

5.4   Iterative Projects: Never-Ending Horror?

Debugging is not an exact science; it's an art form. Unfortunately, the patrons and practitioners of this art both live in a very complex universe, as the following will illustrate:

Suicide Squirrel Syndrome

Sometimes, a bad day in the lab can become a *really* bad day in the lab, and then an *amazingly* bad day in the lab. Picture it: You're working to an (expired) deadline. You observe a problem with your system, and you form a theory about its causes. You assume that a few things are impossible, and spend a few hours investigating the remaining possible causes, frustration building all the while. Just when you're about to tear your hair out, you notice some small piece of evidence proving that your original cause-and-effect theory was wrong, and leading you to a second underlying problem. The theorize-investigate-hit a brick wall cycle then repeats itself.

A peculiarly apposite example of this happened to a colleague of mine a few months ago. He was debugging a piece of code known to be infested with problems. After removing some particularly egregious coding error, he tested the resulting build and found that it was completely nonfunctional. The engineer spent about a day and a half debugging code; breakpointing at different places, checking flags and ADC readings, and so forth. The code in question involved reading the output of an analog circuit and performing some timing calculations (among other things). After making some measurements, he concluded that the reason for the code's weird post-bugfix behavior was that the analog circuit wasn't working quite right. (He had previously assumed that this was impossible, because the unit in question was from production, not a prototype—it was fully tested in the factory. How could it be malfunctioning?) Measuring some voltages on the board, he soon found that the power rails were rather low. After spending another half-day poking around the board looking for a problem in power regulation, or perhaps a miswired part pulling down the rail (note the implicit assumption here as well!), he realized that the 5V DC input to the board was low. Tracing back, it transpired that the lab power supply was reading normal on the front-panel (digital) output meter but the actual output was low. The root cause of this was that on that one particular outlet, the actual AC line voltage was only about 86V! A combination of extremely unusual root causes and assumptions that would normally be valid (but weren't in this case) wasted more than two entire days of effort.

In Australia, power glitches and transformer blowouts are fairly often caused by possums shorting out power wires while climbing among the insulators. Note that the possum in question is typically *Trichosurus vulpecula vulpecula*; this is an arboreal animal, not very much like the *Didelphis marsupialis* opossum that Americans associate with the word "possum." The closest American animal, in terms of climbing abilities and propensity for exploration, would probably be a squirrel. Hence, when a situation like this appears to be developing, we now refer to the probability that a suicide squirrel attack is in progress.

Customers—internal or external—practically never tell engineers what they really want. This is a universal truth that applies equally to large companies, small companies and one-person consulting businesses. The problem is exacerbated by the fact that every engineering decision has implications that are not always obvious even to the engineer responsible for the design. These implications are almost certainly totally opaque to the customer. It's often very hard merely to explain the nature of these problems to nontechnical customers; tragically, you can often proceed quite a long distance down a given design path before the implications of that path become apparent to the customer, at which time it's too late.

You'll begin to notice difficulties of this sort from the very outset of a project, during the specifications development phase. It's almost completely unheard of for a customer to call you and say "I need a widget designed; here are all my requirements." The way small consulting projects generally begin is that the customer calls you to describe a problem that they need to solve with custom hardware and/or software. The problem statement will be presented in terms relevant to the customer's field of business, *not* necessarily to embedded engineering. At this point, you're facing a choice: educate the customer about electronics and embedded programming so they can provide you with a detailed specification, or educate yourself about the customer's field of business so that you can develop such a specification for yourself; pragmatism dictates that the latter path is preferable.

Since it's difficult to absorb all the details of an industry in a few moments, this process is generally iterative. Your customer will state the problem, and you'll respond with a preliminary suggestion as to how the desired solution can be implemented. In this response, you should include as much detail as possible covering issues such as:

- Development time

- Bill-of-materials cost,

- A list of what can and cannot be achieved with the system in question, since the customer probably gave you an endless laundry list of desired features

- Special issues the customer might not have considered

Regulatory issues frequently fall into the last category; for instance, I've been asked to develop software for gambling machines, despite the fact that it is not legal to own these machines[6] in my state nor the state where the other party resided. On a more practical note, almost any electronic appliance sold in the United States needs to comply with FCC Part 15 rules on RF emissions; customers usually don't think of this, and most consultants (myself included) don't have the equipment to perform official FCC testing, so the contract for something that will eventually be a consumer appliance has to include wording to the effect that it's the customer's responsibility to have the product tested and certified, and your only responsibility is to provide engineering support during the test process. (Remember to budget all the possible issues into your quotation, by the way.)

---

[6]   Exceptions apply, but they weren't relevant in this case.

Delivering the initial specification document will trigger another round of discussion with the customer, and another round of specifications, and this process can go on for quite a while. Hopefully, you are spiraling in and converging on a solution that's acceptable to the customer, but this doesn't always happen. Be very aware of this possibility; you can waste a terrible amount of time going backwards and forwards researching all the issues surrounding a project like this, and that's time that would be better spent on more profitable work. (It's at moments like these that you can really appreciate the people in a big company whose full-time job it is to work on negotiating these details with suppliers and customers, so that engineering gets at least a fairly accurate specification with most of this iterative stuff filtered out of it.)

Once you nail down the specification and start working, unfortunately you'll find that customers almost invariably start to throw in additional last-minute changes. It's a basic and practically inviolate rule of engineering—hardware or software—that *the later these changes are introduced, the more expensive they are to implement.* You really can't do anything about this problem except to keep the customer informed, as accurately as possible, what these change requests are going to cost in time and money terms. (Experience is a great help here.) Unfortunately it takes a lot of discipline to prevent these changes from interfering with the overall integrity of the system design. There's a horrible temptation to patch the new feature in quickly in software or hardware, and this almost invariably results in a loss of overall design integrity.

Be very cautious, then, about these last-minute changes. Merely by virtue of the fact that they are indeed last-minute, they will not have received the same scrutiny as all the other features that you and the customer discussed before "finalizing" the specification. It's here that you're going to find the lurking bear traps; the things that can't quite be done in software, the feature that doesn't quite match the way the rest of the system is supposed to work, and the functionality that's going to be just a little bit more than your hardware platform can handle. Changes that are truly minor—switching the red LED for a green one, or changing the default volume setting on a device with audio output—are not a cause for worry, but more complex changes should, if necessary, trigger a requote for the entire project, and development of an entirely new specification and delivery schedule. Patching a project just as it's going out the door is a recipe for expensive disasters.

5.5   Pricing Your Services Appropriately

It is a capital mistake—frequently made, and extraordinarily hard to undo—to underprice your services or to set payment terms that are excessively liberal. If you overprice yourself, you will simply not attract many clients. You can observe this fact and start to reduce your pricing as appropriate. If you underprice yourself, however, you'll get into a much worse situation. You'll start to build relationships with people based on your lowball price, you'll start to rely on the income stream coming from this work, and if you try to adjust your pricing upward at a later date, you may find that the resulting drop-off in business affects your financial survival. Note that I don't mean by this that your customers will say you're not worth paying extra. What I'm warning you against is that if you underprice yourself, you may be building your future financial stability on a raft of customers who simply can't afford industry standard rates for your expertise.

The most rational way to judge how much to charge for your time—particularly if you're moonlighting—is by analyzing the opportunity cost of doing freelance work. The opportunity cost of an activity is a piece of economics jargon meaning the potential returns foregone by choosing that activity instead of the most profitable alternative. To give an example, let's say you have been offered the chance to give an hour-long lecture one evening, in return for a speaker's fee of $5,000. You estimate that the lecture will take you four hours to prepare, and that there will be perhaps two more hours of travel to and from the venue, and an hour of question time involved.

If you don't give the lecture, let's say your next most profitable activity would be working on a contract job at $100 per hour. Giving the lecture would consume eight of these billable hours, so the opportunity cost of giving the lecture is $800 (assuming that you actually have contract work lined up, that you would have to decline if you give the lecture).

The situation isn't really as clear-cut as I have presented it, by the way. A creative accountant looking at the same situation would undoubtedly point out that giving a lecture counts as advertising your services, so the lecture really earns you more than $5,000. Assessing how much this sort of advertising or goodwill might be worth in dollar terms is a specialized art; experience will teach you what business you can expect to come out of a session of this sort.

There's another very simple heuristic for working out what your hourly rate should be (short of asking colleagues working in the same arena, of course): Study a few salary research engines such as salary.com to learn what the going day-job rates are in your geographical area and field of expertise. Calculate the effective hourly rate from the quoted annual baseline salary (don't include stock options, 401(k), medical plans and so on), and double it. This is a surprisingly useful rule of thumb; doubling the nominal hourly rate covers self-employment tax, health insurance, personal liability insurance (if you carry it), utilities and all the other miscellaneous expenses that a day-job employer would normally carry for you.

So much for working out your hourly rate. Projects that require you to buy special tools or unique, expensive parts—or even simply projects that require you to ship the customer a prototype built from parts you normally keep on your lab shelves—are a little more complex. One method of handling contract work of this type is to assemble a bill of materials up-front and send it to the customer, along with a suggested supplier list. The customer should then order the parts for delivery directly to you. The benefit of this system is that you have no out-of-pocket expenses, and the customer has a very clear understanding that you're not marking up any of the parts—you simply bill for hours of work.

There are two major downsides to this approach. Very few nontrivial projects are fortunate enough go from schematic to fully functional prototype in one step, especially when the underlying specification is changing. One practically always has to tune the device—tweak a resistor value here, choose a meatier transistor there, or switch to a less noisy op-amp elsewhere. As a result, you find yourself held up waiting for the customer to approve additional expenses for the prototypes and get the parts in-house. You also risk annoying the customer, who might not understand that some aspects of engineering are a little speculative; some things are hard to develop analytically, so a little trial-and-error can creep in.

For this reason, it's a better idea to charge the raw materials costs to the customer on project completion (or at a major milestone) and simply keep the parts you're likely to require on hand in your lab, replenishing when necessary. In order to avoid being out of pocket, you need to structure your terms in such a way that the initial project-start payment will cover all the anticipated raw materials costs, as well as whatever you need to pay the bills while you're working on the project. A very typical sort of arrangement for smaller projects would be 50% of estimated costs up-front (before commencing work) with the remainder due net 30 after project completion. For a larger project, you might split the billing up further; for example a third of estimated costs up-front, the next third due when the first prototype is delivered, and the remainder due net 30 after delivering the final software and/or hardware.

Another cost to you is the time you spend developing software and hardware. It is clearly going to be a big savings to you if you can re-use design elements from project A in project B. However, if these two projects are being carried out for different customers, you might have a potential conflict. Rather than just ignore this matter and reuse code and circuit blocks from your old designs, I suggest you specifically acknowledge this matter by stating in your terms and conditions that the customer is purchasing a *nonexclusive* right to use the design. Exactly how you word this depends on how much you normally release to the customer; my preference (and customers' preference too, if they know what they're doing) is to release all source materials including sourcecode, precompiled binary, schematics, simulation models (if any), Gerbers to make the PCB, a sample bill of materials with orderable part numbers, and anything else that another person would require to take over the project from you. Some people, however, prefer to release only the binary files and Gerbers; that way, the customer has to come back to you if they want any changes made. I'd rather not lock people in this way; I'm so busy that I can't guarantee I'll have time to look at the next version of a project. It's more ethical of me to release everything and give the customer the option of going somewhere else for maintenance.

While we're on the subject of billing and quoting, a word to the wise: Charge by the hour. Avoid the temptation to quote a fixed dollar amount on a project basis. Even with the best customer in the world, you'll find yourself being nibbled to death with little feature requests if you quote a blanket cost for the whole project. Remember that your time costs money—if you're working to a fixed quote, then any additional feature request from the customer either requires requoting or you need to accept that working on that additional feature is effectively a cash gift from you to the customer. Project quotes work well only when you have a full specification up-front. (I've never seen one of these, either in industry or in private contract work, by the way; every specification from which I've ever built a product was changed by the maintainer of that specification during the implementation process.) The only exception to this general rule is that it's not usual to charge for the hours spent on a quotation. I would, however, advise you to charge for quoting on a project that involves significant research; the exact cutoff point for what you consider "significant" depends on your own preferences. As a rule of thumb, if you're spending more than two to three hours on a quotation, or if you need to call suppliers or component manufacturers to research the feasibility of the project, you should probably charge for the quotation. Otherwise you run a risk that the customer will look at your quote and then give it to someone else to implement. That other person will be riding the research effort you put into preparing the quote; in other words, you're paying your own competitors!

5.6   Establishing Your Own Working Best Practices

I cannot possibly overstress the need for freelance one-person engineering consultancies to be both ethical and well-organized. There are a huge number of things that need to be managed in such a business, and if you don't pay due care to these issues, you will present a very chaotic appearance to the outside world. This is not something that is calculated to draw in future customers. It is imperative, therefore, that you create a system for carrying out your business, and that you stick to it consistently. In fact, it's preferable to have a primitive system that you use consistently than an advanced system that you don't often follow; given the potential complexity of migrating a business from one operating method to another, it's also usually preferable to stick with an existing system until it is absolutely impossible to delay migrating to something new. You've got engineering to do; you don't have time to be re-sorting files, scanning documents and so forth. On the other hand, if you just break off from one system and start a new one, you'll have a lot of annoyance (at least in the first year or so), because you'll constantly be looking for data that are not in the new system.

To begin with the easy part: As I intimated earlier, there's an unavoidable slug of bookkeeping that you'll need to do if you're running your own business. This paperwork is like a bacterial colony; it grows slowly at first, but if left unrestrained, exponential growth will bring it to a really terrifying size in short order. The only way you can keep this monster reined in is to schedule time to work on it, preferably daily. I like to reserve between half an hour to an hour at the end of the day to fire up my financial management software and enter any bills paid or checks received. I then file any paperwork—invoices, correspondence, and so forth—in hanging files. Finally, I update my journal with notes about things achieved during the day and what I expect to get done tomorrow.

On the topic of journals, it's a very prudent idea to buy laboratory notebooks and keep a running diary of what you're doing from day to day. There are all sorts of occasions where such a document can be useful. For instance, you may discover something patentable, and these laboratory notebooks will be excellent proof of the date you made this discovery, should the matter ever come to court.[7] If there's a problem with one of your fielded projects, you might conceivably also have to document and justify how much time you spent on a particular client's project, and what specific analyses you performed while designing the device. On a less legalistic note, you'll find it very handy to be able to look back over the years and look up old designs, ideas and notes you made. Be liberal in your use of sketches and descriptive text, and don't be afraid to include jottings such as mathematical notes and so forth.

Another type of document you'll need to create and file is project specifications. You need to have a consistent method of filing your projects so that all required data (sourcecode, build tools, and so on) are together in one place. That way you can deliver copies of the entire project to the customer, should this be necessary. Having all these materials filed coherently also helps you look up old designs if you need to reuse something. I prefer to do this archiving electronically; I number each project I work on (personal projects included) and create a top-level directory with the project number. Under that directory, I create a separate subdirectory for every version of the software and hardware released to the customer (I don't generally archive internal revisions, only revisions that left my hands and went to someone else). I also archive the complete set of tools used to build the software, along with instructions on how to build or install them, and all the datasheets for the parts used in the design. (Remember that when a

---

[7]   There's rather more to it than what I've written here, if you're worried about patentability issues. For an example of the sorts of things you will need to do in order to establish these books as standalone legal documents, visit the Scientific Notebook Company's instructions page at *<http://www.snco.com/ instruction.htm>*. For definitive answers, you should consult an intellectual property lawyer.

part goes obsolete, the datasheet vanishes off the manufacturers' website, and since nobody creates printed databooks any more, it can be very hard to locate datasheets for long-obsolete devices.) When the project is complete, I burn the project folder to a CD or DVD and send it to the customer; this is how I define project completion, in fact.

Managing your stocks of components and other supplies (solder, wire, and so forth) in the face of constant vendor reshuffles is another annoying task. Most major companies manage this sort of thing by maintaining a database of in-house part numbers for every part they have ever used. Each part number has a specification, and a number of approved vendors attached to it. When you order, say, a reel of 10K 0603 5% 1/16W surface-mount thick-film resistors, it might be coming from any of three different vendors, but when it arrives, you dump it in a bin—literally and from an administrative standpoint—with parts from all the other approved vendors. This sort of system ensures that you can always order parts "a la carte" without needing to worry about changing vendor part numbers; your schematics only show in-house part numbers. Unfortunately, it's usually too much work for the average one-person shop to maintain such a database, so you just have to be prepared to do quite a bit of research when you go to order a complete bill of materials for a project. Some distributors—Mouser, for instance—have handy website features that allow you to upload a bill of materials for a project and save it on their site. Next time you need to build ten of that project, you can just look up that BOM on the distributor's website, say you need ten sets, and the order will be automatically prepopulated for you.

Part of your best practices should also involve a plan for keeping your customers happy with your progress. Cargo carriers such as UPS and FedEx determined long ago that it is a key selling point to allow customers visibility into the parcel delivery process. As a rule, it's a bad thing if a New York-bound parcel you're carrying has been accidentally misdirected to Tecumseh, Alabama. However, if this unfortunate circumstance should occur, it is preferable that your customer should know where the article is and the expected delivery delay rather than simply having the parcel show up a week late with nary a word of explanation. The engineering analogy is that your customers will be happier with you the more visibility they have into the process. Exactly what this means depends on the size and nature of the project. At the very least, I'd make sure you email the customer each time a major event occurs; arrival of the first PCBs, assembly of the first prototype, and so forth. Once you've shipped working hardware to the customer, it's also good to keep supplying them with interim versions of the firmware as you develop it; that way, they can help you test it and any design issues in the user interface can be worked out during the development cycle.

On a closely related note, it is critically important to customer satisfaction that your clients be able to contact you when they need to. This can present some challenges if you're working a day job in addition to carrying out contract work. I've experimented with various forms of electronic communication, and I find that the best solution right now is a BlackBerry device. There are various other appliances that offer cellphone service and email access, but I keep coming back to the BlackBerry because it's fast, widely supported and easy to use. With this type of device, you can keep in touch via email and view some critical attachment types (PDFs and Microsoft Word documents, for instance). The appliance has a long standby life and it's simultaneously physically small, robust and easy to use. Although most regular cellphones can now access email, it's *very* irksome to read and reply to normal text emails with the tiny screen and weird keyboard entry method of a normal cellphone.

Speaking of phones, remember that it's good for you to answer the phone when a customer calls, but it's much better for you to pick up the phone and initiate contact proactively. If there's something the customer needs to know—even bad news—it's much better for the business relationship if you make the phone call as soon as you find out, rather than waiting for the customer to call you.

5.7   More Than a Handshake: The Importance of Contracts

It's a common misconception that you can do business with friends and family without needing any formal documentation of the work. Many people also feel that insisting on contracts, due dates, and specific terms of sale will imply a lack of trust on one or both sides, and undermine personal relationships with the aforementioned friends and family.

This attitude is about as far from reality as you could possibly get. If I could brand just one succinct maxim onto your brain in this section, I'd like it to be this: "Friends don't let friends do vague projects." Contracts ***do not*** create acrimony and ill-feeling—they prevent it! Frankly, if you think your customer is going to cheat you in the contract, or if you're looking for ways to cheat them in writing, you shouldn't be doing this job anyway. Furthermore, if you can't come to a written agreement with somebody, this is a priori evidence that you will not be able to reach—*and adhere to*—a verbal agreement either. You might shake hands and agree over coffee at the start of the project, but it is absolutely certain that you will be arguing come the end of the project.

The first fundamental reason why you need to have a contract in place is to avoid causing surprises to either party. If you have a contract that says "Party A shall pay next-day air shipping to send the first working prototype to Mount Everest," Party A knows this in advance and can plan for it. If a detail like this isn't specified up front, Party A may expect the finished widget to arrive on-site in a week, with no extra money payable—and Party B may be expecting to send it by four-week slow boat at Party A's expense.

The second fundamental reason why you absolutely must have a contract is because the contract—probably by way of an engineering specification—defines when the project is finished, both in terms of time and deliverables. This has all sorts of important ramifications:

- Your client knows exactly what they are going to receive for their money. This is especially important if you've quoted on a whole-project basis (though I don't recommend this; see Section 5.5).

- You and your client both understand what stages trigger responsibilities on either end.

- You have a clear idea—assuming you can meet your due dates, of course!— as to when you will be free to take the next job from someone else.

Speaking of contracts, you'll be asked from time to time to sign nondisclosure agreements (NDAs). Not infrequently, the person asking you to sign the NDA wants you to do so before divulging anything at all about the item it covers. My blanket advice to you is not even to read the NDA, much less sign it, unless you are already interested in the project. Some people—on both sides of the client/contractor fence—treat NDAs as a kind of initial courtship ritual in the process of establishing a business relationship, with no real meaning and no risk on either side.

Make no mistake, though: executing an NDA for a project about which you know absolutely nothing is potentially very dangerous. This is doubly true if you are already bound by NDAs to other clients, or (even worse!) to your day job. Consider what might happen if you sign the NDA, and then the customer tells you about the project—which just happens to be in direct competition with your day job. Even if your day job allows moonlighting, it would be exceedingly unethical to work on both projects.

In a similar vein, consider what might happen if you sign an NDA for (and receive disclosures concerning) a project that turns out to be in competition with, or even just related to, contract work you're doing for someone else under a separate NDA. Even if you never work on this second project, you're still in trouble. When the first project is completed, the person who wanted to hire you for the second project can sue both you and the company that hired you for pilfering trade secrets. In short, therefore, signing NDAs willy nilly is digging yourself into a legal hole, getting out of which can be very expensive.

Of course, simply having a client ask you for paperwork of this sort is no reason to be discourteous, and it's not necessarily a show-stopper to working on the project, either. If I'm asked to sign an NDA up-front, my standard response is that my policy is not to sign NDAs for any project except as part of the quotation process (since this process is the first formal transaction in an engineering relationship that involves exchanging of detailed specifications). I then encourage the other party to provide further information on the project so I can decide if I'm competent to carry it out and interested in doing it. If you follow this method, be sure to stipulate that the other person should not disclose any proprietary information to you. The test I ask people to apply is: don't tell me anything you wouldn't tell a stranger in a bar.

As a side note, I've worked in companies that deal with a lot of external inventors. The toy industry, for example, garners a lot of its original ideas this way. To avoid both the Scylla of letting a good idea get away to the competition and the Charybdis[8] of being sued by an inventor who visits, shows off something the company is already working on, then starts a lawsuit when the company's own, independently developed product comes onto the market, at least a couple of large toy companies simply buy up ideas in bulk and archive them.

My own experience with external inventors is that the ones who insist on draconian agreements up-front are usually inexperienced inventors just getting into the field; I've almost never seen an inspiring idea presented under these conditions. Some of these people in fact are simply hobbyists who have been lured into one of those widely advertised "invention promotion" scams. You can read a great deal about these scams on the FTC's website at *<http://www.ftc.gov/bcp/conline/pubs/services/invent.htm>*, and you might also want to look at the National Inventor Fraud Center, Inc., at *<http://www.inventorfraud.com/>*.

Briefly, there are many companies that advertise the ability to help you turn your idea into a marketable invention and make millions.[9] What they actually do, of course, is employ the usual sorts of con artist lures to inveigle you into parting with more and more of your money.

---

[8] Mythical creatures from Homer's *Odyssey*; they guarded the Straits of Messina. Scylla was a six-headed monster; ships that sailed too close would lose six of their crew. Charybdis created a whirlpool and a fountain three times a day; ships that came too close to her risked being destroyed utterly. Both creatures were originally beautiful nymphs; Charybdis was sentenced to her fate by Zeus for stealing sheep, and Scylla was enchanted by the wicked witch Circe.

[9] Alpaca farming was the big home-industry scam during the last century. Invention promotion may be the corresponding cottage industry scam of the twenty-first century.

6

Working for a Small Company

6.1 Analyze Your Goals: Benefits and Downsides of the Small Company

In this chapter, I'll be talking about life in a "small" company. By this, I mean a corporation with a single place of business (possibly with a few work-from-home offices attached to it), which is not usually dominant in the markets where it sells its goods or services. There also exists a special class of small engineering business where the second criterion doesn't apply, viz. the company (often a startup) with a truly innovative product, a niche product for vertical markets, or a product consisting primarily of IP (intellectual property) that is heavily protected by patents and copyrights. This type of company is "dominant" in its market by default, since it is the only significant player in that market.

As you'd expect, life in a small company has both advantages and disadvantages that are more or less in the middle between freelance work and working for a big corporation. Whether it's right for you or not depends mostly on individual preference. Working in a small company is, at the very least, excellent preparation for working in a larger company. Possibly the biggest advantage (some people would feel that it's a disadvantage) of working in the small-company environment is that every member of the team has cross-functional abilities. The great thing about this is that you'll get to learn a lot about how to run a business, and you'll also be exposed to challenges at a variety of levels in the organization. At the same time, however, you'll have other people to lean on for the "infrastructure" work—paying bills, ordering supplies, and so on—and a bigger financial cushion to sit on than you would otherwise have as a freelance consultant.

You'll also be happy to learn that the procedural workload at a small company is at the absolute minimum level you'll find in an engineering job. Freelancers need to do a fairly large amount of procedural stuff because there's nobody else to keep the lights burning and bills paid, and engineers at big companies need to spend even more time on procedures because there are company policies that force them to. Working at a small company, you avoid both these pressures; it's entirely possible in many small companies for an engineer to get away with virtually no procedural workload at all.

Whether this is a Good Thing or not is an interesting question; engineers and professors both complain constantly about the "code cowboys" who work, unsupervised, at small companies. Regardless of what you can get away with in terms of paperwork, design reviews and similar irritations, it's important to deliver an objectively high-quality product at the end. While it may be the case that your management will reward you based on delivering a product that boots up and works, regardless of how hairy the design might be under the hood, remember that every project you work on is a potential addition to your portfolio, to be shown to another employer. Also contemplate for a moment that the highest goal of many small companies is to develop some highly saleable technology and be acquired by a large company. When that big integration day arrives, your design will be subject to scrutiny by real engineers evaluating the value of your company. If you did an amateurish job, you'll not only look bad, but you might jeopardize the sale. Since acquisitions like this frequently[1] result in large payouts to the engineers concerned, you're doing yourself a big disservice.

Small companies offer a broad spectrum of salary ranges and nonsalary compensation. In general, you can expect a reasonable salary range and an employer-sponsored healthcare plan. Many small companies also offer profit-sharing schemes and a retirement plan of some kind. Beyond that, the benefits start to get fuzzy. You shouldn't expect to find tuition reimbursement, a pension plan,[2] employee discount program and so on at a small company, although

---

[1]   Not always, of course. After you've been with the company for a while, try to secure some equity. This is your surest path to a big payout come acquisition day.

[2]   Note the difference between a pension plan (this is a specified payment awarded to you on retirement based on years of tenure) and a retirement plan (this is a fixed balance, accrued during your working years, and distributed according to your preferences after you reach retirement age).

some companies do offer this sort of incentive, of course. The good news is that small companies are much more likely than large companies to offer you a cash bonus at the end of the year; average raises are also usually more directly tied to profits. The flip side of this can be quite bleak—since your salary is directly linked to the company's ability to make sales, and the cash cushion is usually fairly small, any more-than-momentary interruption in sales can mean you're not getting a paycheck.

The strengths of a small organization lie in rapid adaptation and immediate flexibility. Because a lot of processes are not formally documented and regimented, employees are free to modify their behavior on a case-by-case basis to match changing circumstances. Besides allowing the company to provide better short-term customer satisfaction, this results in a feeling of self-empowerment for the employee which can be very good for morale. The negative side is that such a high level of flexibility introduces an unavoidable level of chaos into the daily operations of the business.

Regardless of this downside, properly handled, the time you spend at a smaller organization can help you develop a solid work ethic and start you on the way to building a network of vendors and colleagues who will work with you in later life. Perhaps most important of all, you'll acquire a broad spectrum of experience in the lifecycle of engineering a product.

6.2   How to Get the Job

While hiring decisions are, of course, individual, there are some general rules that you'll find useful when looking for a job in a small company. First, consider that the person who is going to read your résumé and interview you is almost certainly one of the principals of the company, with sole responsibility for hiring you. Contrast this against the large-company situation, where the gatekeeper who first reads and potentially rejects your résumé is a computer program, and the person who interviews you is merely going to make an upward recommendation to hire you. As a closely related fact, the person who interviews you will have a direct interest in the value proposition (what will this person earn me versus what I will have to pay them); this is definitely not always the case in a large company.

Getting to the point of being offered an interview, then, requires demonstrating (in your résumé and cover letter) a track record of concrete work achieved, preferably with rapid and impressive results. The easiest way to do this is to check off specific projects you've accomplished. Unless there is some particularly amazing personal project you've worked on, you should give precedence to commercial projects, and (if applicable) emphasize what cross-functional responsibilities you had in the project. For example, if you were the lead engineer on a widget and your responsibilities included developing the code, supervising another engineer to develop the hardware, and working with a Far East factory to get the PCB laid out and the product built and tested properly, this should all be in your cover letter. If you can include details about the development time (time to working prototype and time to release), it's a good idea to do so. By the way, my suggestion is to put pointers to the really employer-specific stuff (projects that relate directly to your target industry, for instance) in your cover letter as well as the résumé. Some guides advise you to keep the cover letter terse and rely on the interviewer reading the entire résumé to learn about your career to date; for the case where the result is to be processed by a real human, I don't agree with this at all. The cover letter is like the first chapter of a book; you need to persuade the reader here that the remainder of the text is of interest. Cover letters have a different importance for the large-company hiring process; see Section 7.2 for more information on this topic. I suspect that people who advise a short "please, thank you" cover letter are influenced heavily by the large-company hiring process.

This part of the hiring process is more difficult, of course, if you're fresh out of school (or if you haven't completed school yet); you won't have any commercial projects under your belt. However, you can still pique an interviewer's interest by showing off personal and school projects that you've worked on. You will set yourself apart from the pack by including evidence that you devoted some thought to the issues that affect real production but are often neglected in personal projects; for example, future availability of the parts you designed in, expansion possibilities, ease of manufacturing, ability to be tested automatically, and bill-of-materials costs. It's also generally less impressive to show a project built out of custom parts bolted to a surplus widget than a project you engineered yourself from the ground up.

Once you actually get an interview, you have a chance to amplify on the materials you submitted in the application. (I'd suggest you also bring some physical samples of your portfolio projects with you, so you can demonstrate them if you wish.) You also have a chance to learn if the person who'll be hiring you is addicted to some specific buzzword technology or design methodology; generally speaking, that's something of a red flag hinting at future micro-management, particularly if the interviewer doesn't have practical experience with the technology in question. It's stressful to work for someone who is going to keep their fingers in the design unnecessarily, and doubly so if the person in question has less experience in the field than you do. On the other hand, if you're really desperate for the job, you may contemplate simply buckling under and going with the manager's view of how things should be done. There is a finite possibility that you will find that the manager was right, and the technology they selected is, in fact, a great fit for the job. In my experience, however, this rarely happens; it is much more likely that your frustration level will steadily build until it outweighs the pleasure of a regular paycheck, at which time you will find yourself looking for another job again.

Competition is (relatively) not very intense for small-company positions. They are often not advertised nationally; you'll find listings in newspapers but not always online. (As a side effect of this, when applying for a small-company job it is unlikely, though definitely NOT impossible, that you will be competing against outsourced workers and high-tech visa holders.)

How do you find these positions, though? You can certainly check the online job listings; many of these are trawled from local newspapers. Unfortunately, though, jobs that are listed online attract a *lot* of low-quality applications. It's simply far too easy for someone browsing a website to click "Send my résumé!" and blast a résumé out to hundreds of people; your application will be in a sea of noise. Therefore, I suggest the very best way to find a small-company job is through a real, live headhunter.[3] The comments I made in Chapter 2 about how to find headhunters are entirely relevant to this process.

The next best method to find a small-company job is networking. If you're at school, ask your professors and course advisers about local businesses that are relevant to your field of study. If you've already been in the industry for a while, talk

---

[3]   A book you might find interesting, if a little dated, is *A Big Splash in a Small Pond: Finding a Great Job in a Small Company* by Linda Resnick (Fireside, January 1994, ISBN 0-6717-9807-3).

to your colleagues; many of them will have friends who run small businesses that might be hiring. Failing that, you can potentially glean some ideas from the local Yellow Pages listings and Chambers of Commerce. As a handy hint, you can look up your local chamber of commerce at *<http://www.chamberofcommerce.com/>*. Many local chambers of commerce maintain positions vacant listings online for their members (usually offered through third-party job posting services such as quietagent.com); however, don't restrict yourself to those postings since many openings, particularly from smaller employers, won't make it into those systems. If you see a company whose profile interests you, contact them directly and ask about potential vacancies.

6.3   Responsibilities and Stresses in a Small Company

During my career to date, I've been involved full time with two very different small companies that went out of business. One, a software company, was driven out of business essentially because technology advances made its primary products irrelevant. They did not develop a strategy for dealing with new competing or complementary features built into consumer operating systems until it was too late (and, to make things worse, the specific product they were offering was a feature that is best implemented by the operating system vendor anyway). The other, a hardware company, fell victim to a series of unwise marketing decisions and unfavorable component price fluctuations; they over-extended themselves and died as a result. Both of these positions were, at times, extremely stressful, but perhaps not for the reasons you might think.

It can be terribly lonely working at a small company. By this I don't mean that your co-workers will shun you, but that you may be the only one who is in a position to make engineering decisions about a project. This can be extremely stressful. In personal projects, you can show the design to your professors or friends, or simply publish them on the Internet, and solicit comments about the design. In a large company, your work is subject to design reviews from people who (at least theoretically) will look it over and spot the stupid mistakes before you commit to copper or ship code to the unsuspecting public. In a small company, you may not have any safety net of this type at all, beyond perhaps discussing the project with some close acquaintances. Couple this with the fact that a small company can't afford very many large, expensive engineering mistakes without folding, and you'll see that an awful lot of responsibility rests on your shoulders. The principal factor that differentiates this from working as a freelancer is that your personal assets are not directly at risk when you're working for a corporation (though the loss of a paycheck can, of course, affect them indirectly). There's still plenty of stress to go around, though.

In a closely related vein, one responsibility you need to think about, from an ethical standpoint, is your replaceability. In a small company, it is very easy to become indispensable. The question is, do you really want to get into this position? Particularly if the company falls on hard times, you may want to explore other opportunities. Depending on where your sense of ethics lies, you may not feel good about simply giving two weeks' notice in the knowledge that the company will not be able to replace you. Technically, it is your manager's job to make sure that you are not irreplaceable; it's their duty to put processes in place that ensure that everything you do is sufficiently well-documented that if you're hit by a bus—or recruited by Microsoft[4]—tomorrow, your niche can be filled by someone new.

In practice, few small-company managers understand the embedded engineering position well enough to know how much documentation is necessary, or what form this documentation should take—and that's assuming they even take time to think about the problem (again, most do not pay much attention to the issue). If you organize your work carefully, it will be that much easier for you to train a replacement, should the time come for you to move on to something else. It will also be that much easier for you to train a subordinate if the company grows and needs to hire someone under you. Sure, you can probably scrape along in many companies by scribbling notes on the back of an envelope and making sourcecode backups on a stack of dusty floppies that you throw into the back of an old file cabinet, but you can measure your competence as an engineer to some degree by how easily the next person to park himself in your seat can pick up your work and continue developing it.

As for the freelancer case, I suggest a structure approach to organizing the deliverables you create in a working day. Of course, your first step should be

---

[4]   Given the choice, many engineers wouldn't pick the path you might expect.

to find out if the company has already published best practices for this sort of thing; if not, develop your own—and don't forget to document them! One day, your manager is going to walk in and say "We need to hire a new engineer. Show me what they'll need to know." Or possibly the manager may say "XYZ Corp is negotiating to buy us. Their lawyers want to go over our intellectual property. Show me what I have to give them." In either case, your life will be much simpler if you can just hand your manager a set of backup CDs or point them to a location on the network and give them a document that will explain how to find what they're looking for. It's simply professional to document what you do and how you do it; your work will be all the better for being consistent, as well.

In a small company—particularly one that's not presently doing very well financially—you'll sometimes have a positively uncomfortable level of visibility into the company's sales performance. In the worst case, you'll find yourself listening for the phones to ring with orders (even if it's not your responsibility to answer them) so you can know whether a paycheck will be forthcoming this week or not. Under these circumstances, it is easy for engineers to get distracted from longterm project development into making quick patches and hacks for specific customers in order to secure a particular sale. On the other hand, however, the good side about the strong sales-to-paycheck link is that engineering is keenly interested in the end-user's perception of the product. This can, in ideal circumstances, lead to extremely usable and well-engineered devices, built with the needs of real customers in mind.

The other enormous stress you'll encounter in a small company is dealing with shortages of manpower, cash and time. This topic is so important, and will color so much of your work effort, that I've broken it out into a separate section; see Section 6.5 for a discussion of the issues involved here.

6.4   Personal Dynamics in Small Companies

This section might seem a little out of place in a book about engineering, but you're going to spend an awful lot of time with your co-workers, and it's helpful to remember that they are live human beings. A few words are appropriate, therefore, about personalities in the small company. When you're a freelance agent, you have relatively little exposure to the vagaries of the human psyche.

Clients are mostly courteous, and if you don't like a particular customer for whatever reason, you can simply choose to be busy whenever they call to offer you a job. The situation is quite different in a small company; you'll see quite a lot of human nature at work!

If you're looking for a job where you can spend your time squirreled away in a cubicle banging on a keyboard and an oscilloscope, and never have to poke your head out to talk to real-meat people, you're actually better off working in a large company. In the small company environment, you absolutely cannot avoid interacting with lots of people. You'll be a required attendee when vendors come by, because you're the only one who can competently evaluate their parts. You'll be pulled into customer meetings because nobody else will be able to answer the technical questions. Above all, you'll be dealing with most or all of your co-workers every day, because in a small company there are no "in-between people" (department heads, and so forth) to act as the interface between different specialties in the company. You will *have* to deal with the mechanical engineer who is designing your housing. You will *have* to deal with the person in purchasing who is buying the parts for your next build. You will *have* to deal with the marketing person who is preparing to dash off for a presentation. Worst of all, perhaps, you will quite likely *have* to deal with end-users of your product, and talk them through problems they're having. I always dread this part of a small-company job—not because I'm such a misanthrope, but because in my experience, very little tech support time is actually spent debugging your own product. Particularly for devices that connect to a computer in some fashion, almost all the time you'll spend on the phone to customers will be wasted explaining how to drive the customer's operating system, router or some other hardware or software product that you're not directly responsible for maintaining; it's a very poor investment of engineering time. Unfortunately, in most small companies, the engineers are the only staff with detailed (usually formal) education on these hi-tech topics.

As a result of all this face time, you'll get to see a spectrum of moods from your managers and other co-workers. Small companies are often compared with families; well, in a close environment like a family, it is easy for a mere irritation to develop into a massive, career-threatening conflict. It sounds glib, but the best policy you can have when tempers flare is to back down courteously before things get out of hand. Even if you have strong views on a topic, remember that you're only at work; you can keep your personal opinion separate from the consensus of your colleagues about what is "right" without sacrificing any sort of personal integrity. (I'm referring here primarily to arguments about technical issues, but the same principle applies to general discussions about the entire gamut of human thinking.)

Small companies are, of course, bound by the same laws about harassment and workplace safety as large companies. The difference is that small companies don't hire people specifically to go around checking for violations of those laws; nor do they usually have published policies on the topic. Hence, people usually don't have the possible repercussions foremost in their mind when they get into an argument.[5] It's not impossible, therefore, that an argument can devolve into a shouting match, or even a physical altercation; needless to say, this is about the worst possible outcome. It even has the potential to haunt the rest of your working life.

Think of it this way: If you're arguing with your manager about something, what are you trying to achieve? They write your paycheck and expect you to do as you're told. If your professional opinion is that what they told you to do is not the best course of action, then your duty is to tell them this, and explain why. Once you've gone on record with your comments, your manager can choose to override your views. That's their prerogative; they're in charge. While it is possible that their decision will affect your financial well-being (if they run the company into the ground pursuing a dead-end technology, for instance), you can console yourself with the thought that your manager is *constantly* making decisions with the same potential impact on your life, and this particular issue you're arguing about is just one tiny part of that. A good manager makes the right decisions most of the time. A bad manager makes a significant number of bad decisions. If you're working for a bad manager, you should be looking for another job already, because the business is almost certainly doomed.

Finally, it doesn't happen all the time (hey, we're talking about *engineers* here, remember!), but occasionally an opportunity for romance presents itself at work. My opinion is that availing yourself of such an opportunity is unequivocally a Bad Thing in a small-company environment. While the potential repercussions

---

[5]   I'm not saying it's necessarily healthier to work in an environment where everyone's constantly looking over their shoulder for the thought police. It is definitely quieter, though, and probably safer.

of a quarrel between yourself and a workplace significant other can be worked around with some effort, it's usually unreasonably hard to do so in a small company. If you work for a large company, you can be moved to another department or at least put under a different manager. In a small company, your only way out may be to leave your job, or persuade the other person to do so; in either case, you're creating an unfavorable impression in the mind of the manager who now has to go out and hire a replacement for the missing person.

6.5   Managing Tightly Limited Resources

Here's a short quotate from *This New Ocean—A History of Project Mercury*, a book about the Mercury capsule program that put the first American in space.[6]

The manufacturer of the RCS [reaction control system—part of the spacecraft's on-orbit maneuvering apparatus], Bell Aerosystems Company, ran its qualification test program from August through October 1960 and reported all phases of the testing satisfactorily completed. Subsequent tests by McDonnell, STG, other NASA engineers, the preflight teams at the Cape, and eventually by the workers on Project Orbit revealed innumerable electrochemical and electromechanical problems in simulated environments that required small changes here and there and eventually everywhere. The thrust chambers, metering orifices, solenoid valves, expulsion bladder, and relief valves each presented developmental flaws that were "solved" more often by improvisations than by scientific redesign. Karl F. Greil, a thermodynamicist who was working for Grand Central Rocket Company in 1960 to perfect the escape pyrotechnology for Mercury, joined STG and its reaction controls test team in 1961 and tried in vain to apply the same perfectionistic standards to this vastly more complicated and inherently less reliable system of moving parts.

This is the irony: the results that counted in Mercury's RCS were due to changes of the screen, heat barrier, and orifices, all of which were made upon simple first thought. On the other hand, the large amount of experimentation on the valve resulted merely in the assurance that nothing needed to be changed so far as valve design was concerned. This irony, that the simple approach did the entire job while the sophisticated approach merely resulted in an "Amen", is indeed worthy of reflection, because it has in store both a risk

---

[6]   This publication is by Loyd S. Swenson Jr. et al. (online edition dated 1996, though it also carries a date of 1989, NASA publication SP-4201). At the time of writing, you can read the full text of this book at *<http://www.hq.nasa.gov/office/pao/History/SP-4201/cover.htm>*.

and a lesson: a lesson because there is so much glamor cast on sophisticated pretense and so much disregard for the profane causes of all kinds of trouble; a risk because the simple remedy which did the job once without ever having become clear just how it really worked, such success without perspiration is likely to remain confined to its own historical case. But having established a precedent, it is bound to seduce us into relying on it, if it is not even bound to become a myth and a dogma.

The small-company engineer is used to running short of almost everything—mostly time. Every engineer with any significant experience at all is familiar with the zen state of deep thought that accompanies their most productive moments. Those are the days when you sit down at your desk and the code and circuits just flow magically out of your fingertips. This time is extremely valuable, and it's rendered almost useless by interruptions. (For some light reading on this topic, refer to Jack Ganssle's Embedded Muse #53, October 23, 2000, and specifically the references he sites—you can find this edition of the newsletter at *<http://www.ganssle.com/tem/tem53.pdf>*.) The occasions when you can reach this peak of achievement are the times when you'll do your best code. Outside these times, you'll almost always be in a rush. This can lead to carelessness, and (less destructively) to a large number of solutions "which did the job once without ever having become clear just how it really worked" as in the preceding quotation. Be aware of when you're creating magic answers like this, and consciously try to avoid the temptation. Reusing existing work is the key to continued productivity, and it's very hard to reuse magical solutions.

Many small companies keep their employees working in open-plan spaces without private offices. As a result, you may need a method of manufacturing your own privacy to avoid interruptions. During crunch times, while working in an environment like this, I obtained a good set of noise-canceling headphones and used them to listen to relaxation sound-effect CDs. (I find that listening to almost any kind of music is a distraction when I'm trying to concentrate deeply; you may feel differently about it. The point is to censor out as much of the random stuff going on in your surroundings as possible, and replace it with something unobtrusive so you can focus on the task at hand.)

Techniques like this will improve the amount you get done between today and the scheduled release date of your hardware or software project, but it won't actually create man-hours to help you get the job done. You can pull late nights and all-nighters, but I generally wouldn't advise it—besides the fact that you're putting yourself under tremendous physical stress working like this, you're also establishing an expectation in the rest of the company that you're the guy who will always be there to burn the midnight oil when it's needed. It isn't worth it, even if you're being paid overtime—which I'm sure you won't be. If there is a very good, *short-term* reason for working late, then by all means do it (though I'd suggest that you tell your manager beforehand that you would like to make up the hours with a bonus vacation day or two after the crunch is over). Very good reasons include a manufacturing problem that's going to sink the company, or a design problem that is causing safety issues in the field. Anything of smaller magnitude than this doesn't cut it. Simply wanting to meet marketing's promised delivery date on a widget isn't a good enough reason unless the customer that's waiting is actually going to cancel their order *and* that cancellation is big enough to affect the company's viability seriously. Burnout is a very real problem—don't let it happen to you.

Because of the constant time pressure, exacerbated by the fact that most of the company is nonengineering and doesn't understand a good design flow, it is excruciatingly difficult to follow a rigorous design process in most small-company environments. As a result of this, as I indicated previously, it's easy to wind up with code and circuits that "just work" without being well-understood—hence the quotation with which I began this section. If you can achieve a documented, standards-compliant design flow in the small-company world, then kudos to you—but you should plan for the probability of a short-circuited design cycle. It often happens that you have no time even to breadboard a circuit before building it in the factory. You may then wind up having a technician trial-and-erroring with resistor and capacitor values on the first run of prototype boards while you work feverishly on the code or other aspects of the hardware. This is of course more haste, less speed—but that is very hard to explain to management.

Working with limited financial resources also significantly affects the way engineers make design decisions. (My first book went into some detail about this topic; you might find it useful reading.) For one random example, you'll probably prefer to design in chips that are in packages you can hand-solder, since this means you can construct your prototypes in-house. However, it closes the door to a whole range of parts that are only shipped in BGA packages and similar leadless beasts.

Having these constrained resources also shifts the value equation for purchasing tools (both hardware and software). For example, my preferred CAD package for some years has been EAGLE, from Cadsoft <*http://www.cadsoftusa.com/*>. This is undoubtedly the best value-for-money PCB CAD package available, and it happens to be an excellent choice for a small company or freelance contractor who doesn't want to buy several tens of thousands of dollars worth of software. However, EAGLE files are not directly interchangeable with the CAD systems used by most Asian CMs and ODMs.[7] This results in the possibility of transcription errors when you send the schematics to an overseas manufacturer; it also increases the setup time for the production line.

Observe that as your target system becomes more and more complex, given that you don't hire any additional engineering staff, it becomes concomitantly less feasible to develop internal system components in-house. The ultimate end of this curve is that for a really difficult high-speed design approximating a desktop PC in complexity, you will be reduced to buying an actual PC as the core of the project, since there will simply not be enough resources in-house to develop a custom solution. In between, you have various intermediate steps, but in general as you get more complex, your system contains more and more black box subassemblies purchased from external vendors. As this "evolution" occurs, your system develops rough edges because the interfaces between all these components aren't exactly matched. Furthermore, your ability to change low-level behavior is significantly restricted; if you buy an 802.11(g) wireless chipset and integrate it onto your own board, you can make it do anything it's electrically capable of doing, but if you buy a module and simply connect it to an existing board, you are restricted to the features offered by the module's firmware.

Finally, on the topic of constrained finances, you should be aware that most semiconductor vendors will (understandably) not pay much attention to small companies.[8] They will be happy to visit on a sales call, but in general you can't expect a great deal of support unless there are special circumstances. As an

---

[7]   Contract Manufacturer and Original Design Manufacturer, respectively. A CM takes your schematic and other engineering data and builds a product for you. An ODM designs a product to your specifications and puts your brand name on it.

[8]   There are exceptions to this. Three vendors that have provided excellent support for my small-company activities are Sharp, Cirrus Logic and Microchip. The support you get will depend on the region you're in and how your activities match up against the vendor's target design demographic.

example of what might constitute special circumstances, if you're lucky enough to ask about a new part that hasn't had any major design wins yet, you can get some extra-special first-time support out of the chip vendor so they can take your widget to trade shows as a demonstration that the part has been designed into real, saleable product. Given that you won't normally get stellar support from the manufacturer, then, you're going to have to rely a lot more on peer support. Usenet is your friend here, but remember that it's a shared resource. If you use it to ask questions about your projects, please remember to "give back" to the community by taking time to read and answer other peoples' questions.

6.6 Task Breakdown: A Typical Week

You might find it instructive to look at this representative breakdown of a week's tasks carried out by an engineer working in a small company. This assumes a 40-hour work week. In a "crunch time," with customers clamoring for updated product, factories churning out a defect rate of 90%, or some other engineering emergency, you might see this number of hours double; you might also have to come in on a weekend. Of course, this little timetable doesn't directly represent anybody's real-world schedule, but it illustrates the sort of percentage of your time you're likely to spend on particular tasks.

- Monday
  - 1 hour – Responding to email from Far East manufacturing.
  - 2 hours – Providing end-user technical support via email and telephone.
  - 4 hours – Combined circuit and software development.
  - 1 hour – Miscellaneous (conversations with vendors, co-workers, etc.). These tasks generally occupy more time on Mondays and Fridays than other days of the week.
- Tuesday
  - 1 hour – Providing end-user technical support.
  - 2 hours – Meeting with customers.

- 4.5 hours – Combined circuit and software development.
- 0.5 hours – Miscellaneous.

- Wednesday
  - 1 hour – Providing end-user technical support.
  - 2 hours – Meeting with vendor.
  - 1 hour – Drafting notes for end-user documentation.
  - 3.5 hours – Combined circuit and software development.
  - 0.5 hours – Miscellaneous.

- Thursday
  - 1 hour – Dealing with Far East manufacturing issues.
  - 1 hour – Strategy meeting with management.
  - 5.5 hours – Combined circuit and software development.
  - 0.5 hours – Miscellaneous.

- Friday
  - 1 hour – Providing end-user technical support.
  - 2 hours – Reviewing PCB layout.
  - 4 hours – Combined circuit and software development.
  - 1 hour – Miscellaneous.

7

Working for a Larger Company

Most large companies have rules that require anyone at a given level of management to have a certain minimum number of direct reports. After some thought, I have realized that this is because there is a staggeringly exact analogy between the task of engineering management and the problem of thermal management on spacecraft. A stationary spacecraft will freeze on its shadow side and boil on its sunward side. A spacecraft wholly in the umbra of some other object needs to have its own source of heat to replace radiated losses. All temperature-sensitive equipment needs to be mounted and monitored with careful thought to internally generated heat. The spacecraft is covered in foils, paints and blankets designed to reject solar radiation. Complex arrangements of fins, heatsinks and sometimes circulated coolant liquids are required to dump internally generated heat, plus whatever solar radiation leaks in, out the night-side of the craft.

Replace the concept of "solar heat" with "unfavorable attention from upper management" and the analogy is clear. A successful manager spins gently at all times, like the Apollo Command and Service Module in lunar coast mode, so that no single surface receives 100% of incident radiation. The attention wattage decreases exponentially as the manager moves further away from upper management.

Direct reports are a manager's heatsinks; they cling to the main body of the craft and increase the surface area available for radiating attention away from the craft. Under some circumstances, they also form the basis of a sublimation cooling system; the coolant absorbs as much heat as possible, then is boiled into space. You want to avoid, at all costs, becoming one of these sublimed coolant particles.

Since incident attention from above is directly proportional to the manager's stature in the company, the heatsink area required to dissipate that anger naturally also has to increase as the manager is promoted. Company policies about the number of direct reports necessary for a manager to hold a specific title reflect the amount of attention that title is required to dissipate.

7.1 Analyze Your Goals: Benefits and Downsides of the Large Company

Roughly half of all Americans work in large companies,[1] and approximately one third of all new jobs created are generated by these large companies. Working for a large corporation is the traditional career path that engineers have been following for years, and it's not a path that is likely to disappear any time soon.

Developing complex electronic systems can be very expensive, requiring costly tools and a significant investment in prototypes and regulatory testing requirements. Furthermore, in today's environment, most new products are built incrementally on existing platforms and protocols, which means (unfortunately) that they're built on other peoples' patents. Barriers to entry are therefore distressingly high. Only large companies have the financial clout to penetrate some of these markets. For this reason, if you want to work on really major projects—the twenty-first century equivalent to the Apollo program, for instance—or mass-market consumer appliances—you'll absolutely have to go to a large company.

This isn't entirely bad news. Large companies offer financial stability (as long as you work in a department that is making good profits for the company, and as long as you're pulling your weight enough to keep your manager happy). You'll also have access to good laboratory facilities and the best equipment, as long as you can persuade your manager that there's a sound business case to purchase them. Finally, the very best benefits (though not necessarily base pay, and certainly not cash bonuses) can generally be found at large companies—401(k) programs, cash payouts for patent disclosures, premium healthcare and dental plans, tuition reimbursement, employee discount plans with other companies, guaranteed vacation time, and more.

---

[1]   This is defined negatively; a "large company" for the purposes of this statistic is any company that isn't covered by the rules of the Small Business Administration.

The major downside to large companies is, commonly, a loss of individual creativity and freedom. Employees in the large-company environment are usually specialized and pigeonholed for efficiency reasons; they are not exposed to cross-functional requirements except when the hear their teammates reporting on other aspects of the project.

Most large companies also have onerous development procedures designed to fit a theoretical general case; in many cases, these procedures have all kinds of inbuilt requirements that are more or less orthogonal to the idea of developing a quality product and releasing it quickly. Design cycles, even for simple products like toys, are typically long in the large-company world (two years or more) compared with the 6–9 month cycle typical in small companies. All this can be very irksome to a creative employee.

Because most large companies sit on significant cash reserves, there is also a distinction between "my" money and "company" money. Employees therefore often become complacent about development costs. (In many cases, engineers ar the "individual contributor" level will not even have enough information to estimate development costs—they can be amazed when they see the numbers.) Time, in particular, becomes an easy commodity to spend. There is also little clear link between the product's palatability to end-users and the engineer's paycheck; as a result, human factors are all too frequently neglected in large-company products. At other times, internal company politics can cripple an otherwise good product; for example, a desire for the cheap home-user version of a product not to compete with some more expensive commercial or industrial version.

Choosing to work at a large corporation is, for many people, a compromise between financial certainty[2] and freedom. Where you sit on this topic is a matter of individual preference; you should simply be aware that it is most unlikely that you will ever find a large-company position that involves the same broad task exposure and opportunity for individual achievement that can be found in the small-company arena.

---

[2] This is a bit of an illusion, of course. Every day, we read of layoffs and closures that affect the employees of large companies.

7.2 How to Get the Job

Hiring a new employee at the engineer level is a significantly expensive and complicated process for a large company. Simply hunting for the right candidate, paying a recruiter's fees, potentially paying for relocation expenses and the other setup costs for the hire (plus, potentially, a signing bonus) can easily add up to $15,000 or even more for an engineering position. It doesn't have to be a high-level position, either—big companies will even spend this much recruiting the right person for a mid-level position.

Partly because of this complication, and partly because of the unavoidable large-corporation love of procedure, landing an engineering job at a large company is a highly structured process, divided into phases. The first phase is simply to get into your target company's human resources system. This is superficially easy—almost every large company has a website with an easily located careers section where you can submit your résumé. It's worth doing this even if you plan to use an alternate route into the company, because the hire process will probably require you to enter your résumé in the standard manner anyway. Once you have your résumé in the system, you'll be in a position to use the application tools on the employer's website (and you'll also probably start receiving automated emails with positions that match some of the keywords in your résumé).

Now, you just need to get someone to read your résumé. This is really difficult if the résumés are pre-filtered by "intelligent" software. You're really not in with much of a chance if you go in through this standard route but you don't exactly match the required qualifications word for word. In Section 2.2, I discussed how it's easiest if you have a recruiter working for you on the inside; this advice is generically true for large-company jobs regardless of your qualifications. As a result, I strongly recommend you concentrate your search through professional headhunters. It doesn't actually *hurt* to apply for random jobs online, of course, but don't be overly disappointed if you don't land the job of your dreams immediately by this means.

I'll reiterate here that the large recruiters are not really recruiters so much as clearinghouses for résumés and jobs. They compare pieces of paper, then mail notification to apparent matches out to the protagonists. It's vastly preferable to work with a real live person who knows the hiring manager personally (and preferably knows your industry as well). You might also get an interview by virtue of networking. For example, if you did an internship at the company, you can call the manager and tell them that you applied for a particular position; that can be enough to get your résumé pulled into the active file for that position.

Having been granted an interview, prepare carefully! Many large companies will administer some kind of test before deciding that you've got what it takes for the position. These tests can range from the difficult to the irritating. Microsoft's interview questions, for instance, are quite legendary throughout the industry (search on Google for "Microsoft interview question" and you'll find a lot of personal websites listing the questions people have been asked in their Microsoft interviews). More traditionally, you will be asked specific technical questions about some aspect of interest to the target company. For example, when I was being telephone-interviewed by a West Coast console game development company, I was asked to describe how I would debug a problem with a piece of embedded filesystem code. When I was interviewing for the position where I currently work, the interviewing manager gave me five minutes to study a moderately complex schematic of a product containing two microcontrollers and various communications hardware, then asked me to describe the function of the circuit as best I could. He then gave me a snippet of sourcecode from the same product and went through the same sort of procedure.

Those tests I just mentioned are very normal sorts of assessment techniques that shouldn't cause you undue worry. The weird and wacky questions, however, can be a real bear to deal with. I'm not sure I would want to work for a company that bases its hire decisions on these "creative" questions—it's a management fad, and not a good one. However, you have to be prepared to deal with them. If you're interviewing with a big name, you should Google for relevant search terms to see if other people are reporting odd or frightening behavior at job interviews with this big company.

I advise you to go into the interview with an idea of what your long-term plans would be if you are accepted to work with the company. In other words, you should be able to state clearly if you want to spend the rest of your life as a frontline engineer, if you aspire to management, or if you want to branch out into the forward-looking research and development side of the company.

The next step, assuming the company likes you, is to receive an offer letter. This will be a physical piece of paper stating your salary and expected start date, which you might not actually receive until the day you start work. Once you've accepted the employment offer, most large American corporations will then require you to take a drug test; continued employment is contingent on passing this test. (Some companies require you to take these tests repeatedly at random intervals; others merely require you to be retested if you're involved in a workplace accident.) Many people feel this to be an unwarrantable invasion of privacy. While I basically agree with that viewpoint, my primary concern about drug testing is that some of the methods used can give false positives.[3] The two main sampling methods used are urinalysis and hair analysis; the former can be analyzed in a couple of different ways, and it is purportedly fairly vulnerable to false positives from various prescription and over-the-counter medications. The hair sample method is supposedly more reliable (and, as a bonus to your employer, it reveals data about a longer period of the recent past than urinalysis). If you have a choice, which some employers do offer you, I'd suggest you go with the hair method. Naturally, I'm assuming you have nothing to hide here—if you do, you're on your own.

7.3 Globalization: Outsourcing and Temporary Worker Visas

This section is aimed most directly at readers in the United States of America, but it is also relevant to European, Australian and other "first-world" engineers. In all of the developed world, there is a growing terror in professional communities that their jobs, formerly believed to be immobile, can be outsourced to foreign engineers who have a much lower cost of living and therefore work for much less. There is a somewhat less acrimonious, but nevertheless perceptible body of ill-will toward companies that employ skilled foreigners using temporary work visas like the H-1B program in place in the United States.

The main reason I included this section in the book you're reading is that I hear a lot of doom and gloom on these topics from embittered engineers.[4]

---

3   I'd feel differently about this if the testing required blood samples. At least the methods currently in use are *reasonably* noninvasive.

4   I also hear a lot of perky nonsense from bean-counting types about how outsourcing lets the really skilled employees back home work on the important forward-looking projects. Nobody who promulgates this idea seriously believes it.

Specifically, when young engineering hopefuls ask about their prospects in public forums, they are often handed a diatribe about how engineering is a dead end, and the student of today should become a lawyer, doctor or nurse instead, since those jobs can't easily be outsourced. (By the way, this is no longer entirely true for lawyers, at least—foreign legal companies are often used, for instance, in intellectual property research and patent filings.)

Note that I'm also writing this chapter from an interesting perspective, since I presently work for a company that has considerable overseas interests in India and other popular offshoring destinations, so my job is more or less permanently at risk of moving to one of those locations. I also originally came to the U.S. as an H-1B skilled worker—and though I'm not one of those any more, it was certainly a very educational experience about U.S. immigration laws.

There are two competing viewpoints about the end result of the offshoring trend. For want of an official term to describe either, I'll call one the Malthusian[5] theory and the other the Utopian theory. Each of these theories really represents a continuum of beliefs, of course, with the extreme Malthusians lying at one end, the extreme Utopians lying at the other, and a considerable spectrum of different opinions in between.

Malthusians predict a gloomy downward spiral. Overseas engineers are cheaper than American or European workers, therefore corporations will move all their design work to offshore facilities. Even those corporations that want to "do the right thing" and employ local talent will be forced by simple price competition to abandon local development or go out of business. There will ultimately be no jobs left for design engineers in the developed world, except for a tiny handful of positions working on military projects critical to national security. In short, engineering will go the way of many industrial jobs. The pool of experienced engineering teachers in America will evaporate, and we will become totally dependent on foreign countries for our engineering needs.

Another common line of Malthusian thought is that personal wealth will seesaw; engineers in foreign climes will get richer, engineers in America will get

---

[5]   From Thomas Malthus (1766–1834), who predicted a doomsday global famine based on overproduction of human offspring and the inability of linearly growing resource supplies (particularly food) to keep up with exponential population growth. His ultimate recommendation was that the family size of the lower classes should be regulated so that these people would not produce more children than they could support. China's one-child policy is a latter-day example of Malthusian planning.

poorer, and eventually they will achieve equilibrium. The corollary to this latter line of thought is that enrollment rates in engineering courses will plummet, because people will see that it's not a profitable career option. In turn, this will lead to a dearth of teachers and again eventually the extinction of the engineering profession in the developed world. A more extreme variant of the same viewpoint is that the economies of America and its trading partners will act like connected vessels of water, and seek a common level; everyone in America will get poorer, while everyone gets richer in China, India, and so on. Some particularly dedicated Socialists regard this as an inevitable and in fact desirable outcome.

The Utopian theory is vaguer, and has several branches. One popular line of thinking says simply that American minds will dream up the "next great thing," whatever that might be, and engineers will stop doing what they do today and start working in the myriad of jobs generated by that "next great thing" while the foreign workers scramble futilely to catch up with this new technology. The other most popular Utopian dream is that employers will eventually find out that offshore workers aren't as cost-effective as they appeared to be, and there will be a huge wave of renationalization of those jobs. Every press release about a failed offshoring venture is seized upon by these people as being the first trickle of an assumed tidal wave.

When all is said and done, there's no such thing as a true utopia, and Malthus himself was empirically wrong (since, among other things, he failed to predict technology's impact on food production, and the fact that population growth declines as the population becomes richer and better-educated). So the real results of offshoring, free-trade agreements and temporary worker visas are almost certainly not going to follow any of the preceding theories directly.

One of the flaws with the Malthusian viewpoint is that it's simply not possible to sustain a world-class engineering community within a country if that community consists only of a thousand (or so) eggheads scattered across the continent working on secret military projects. Your military geniuses are supposed to be the cream of the crop—if they comprise the entire crop, then you have a rather large problem, because you're essentially telling your tertiary institutions "Turn us out one hundred guaranteed geniuses per year, with a zero defect rate." Apart from anything else, who's going to teach these proto-geniuses and how are they going to ensure that supply and demand are balanced? Keeping a program like this running would require unbelievably tight control and good organization from a set of government entities that are presently unable to ensure even that a randomly selected nondisabled adult is sufficiently literate to read the instructions on a microwave dinner.[6]

As far as the issue of salary equilibrium across international borders goes, in December 2004 the IEEE-USA published a frightening press release[7] entitled "Incomes of Technical Professionals Decline, IEEE-USA Salary Survey Reveals." This article is representative of materials generated by the anti-foreign-worker lobby (if you'll pardon the rather coarse phrase; I synthesized it in order to cover antipathy to both offshoring and temporary worker visas).

The problem with statistics like this is that the numbers are redacted to a point where they admit of many possible explanations. To take just the examples that spring immediately to mind:

- Income wasn't broken down exactly by source; it was simply limited to "primary sources," which were defined as "base pay, plus any self-employment income, commissions or bonuses." There is considerable wiggle room here—perhaps engineers are spending more time at home instead of moonlighting. Or perhaps there has been a hiring explosion, and people who were formerly deriving their income from self-employment have moved into full-time jobs. Cash-in-hand income is lower when you move to a full-time job, because the employer includes benefits such as healthcare, 401(k) contributions, and also because the amount of Social Security and Medicare tax a regular employee pays is half what a self-employed person pays (your employer covers the other half; see Chapter 5).

- There was no information given about experience levels versus income. It's entirely plausible, given only the bare medians to compare, that what we're actually seeing is a huge wave of new professionals entering the job market. As entry-level employees in a glut year, they would obviously have lower starting salaries than their experienced counterparts, thereby pulling down the median. It's not obvious why this should be a portent of doom.

---

[6]   An oft-cited statistic from "Adult Literacy in America," U.S. Department of Education, 1993: In the United States, 40 million adults are functionally illiterate (i.e., cannot read above the third-grade level). Among other things, this means that they are unable to read the directions on a medicine bottle or microwave dinner.

[7]   The full text is available on the Web at <*http://www.ieeeusa.org/communications/releases/2004/122204pr.asp*>.

- IEEE-USA's definition of technical professionals for the purpose of this article includes "electrical and electronics engineers, computer hardware and software engineers, and computer scientists and system analysts, among others." This is very broad.

Offshoring is not necessarily the instant and long-term cost saver it has been advertised to be, either. Rough, anecdotal information I have been given indicates that in the late 1990s, the cost of hiring offshore engineers in India was a tenth the cost of "equivalent" talent in the United States. However, as of early 2005, this ratio has narrowed to approximately one third the cost.[8] While it's possible that some of this is due to earnings declining in the United States (although frankly I'm not even slightly convinced that this is really happening), most of it is due to the fact that the top talent in India is sought-after and highly mobile—and those offshore engineers are jockeying around between jobs to get better salaries, just like their counterparts in the United States. In fact, India is now trying to compete harder in order to avoid losing work to still cheaper centers of skilled labor, such as China.

What of temporary skilled worker visas? There are numerous lobby groups on the Internet denouncing the H-1B[9] program as spelling the death of American engineering (<*http://www.zazona.com/*> is a good example). Again, it's not easy to ascertain just what effect these programs have on the average local worker.

Before we go any further, here's a thumbnail sketch of how the H-1B visa process is structured.[10] Please note that this is *not* immigration advice by any stretch of the imagination; I'm just describing the process in the broadest possible terms so you understand roughly how it works. Needless to say, there are abstruse legal and bureaucratic complications possible at every step along the way.

- The employer decides, on whatever criteria, that a particular foreign person is an essential hire.

---

[8] It's very difficult to get quotable numbers on this, hence the hand-waving. Contract engineering companies don't break down their fees in such a way as to permit easy comparisons.

[9] This is the visa type that allows "highly qualified" aliens to work in the United States.

[10] A very good reference on the H-1B visa process, along with some discussion of how it might be used to depress local wages, can be found in "The Bottom of the Pay Scale: Wages for H-1B Computer Programmers" <*http://www.cis.org/articles/2005/back1305.html*>.

- The employer advertises in local media to obtain a U.S. person with equivalent skills.

- If no such person is forthcoming, the employer can file a labor condition application (LCA) as the first step toward sponsoring the foreign worker. Part of this application process involves demonstrating to the Department of Labor that the employee will be paid at least 95% of the prevailing rate for workers of this type in the area where they will be working.

- Once the LCA is approved, the company then files a petition to sponsor the temporary foreign worker.

- Once the petition is granted, the worker can apply for the actual visa.

- The visa application process involves showing that the applicant has at least a bachelor's degree, or equivalent work experience. There are third-party companies that evaluate these credentials in accordance with USCIS[11] rules. Observe that simply lacking a bachelors' degree is not automatic grounds for rejection; the general rule of thumb being four years' work experience counting for one year of education.

- If the employee is fired, or the employing company goes out of business, the employee cannot work for another employer without first filing a new H-1B application or other request for legal employment status.

- An H-1B visa is good for three years, and it can be renewed once (for another three years). After this time, the H-1B holder must apply for permanent resident status (a "green card"). They are permitted to live and work in the United States while the permanent residency application is pending.

- The H-1B visa allows re-entry to the United States; the holder is able to travel outside the country and return while the visa is valid.

- H-1B visas are now somewhat portable; if you're working as an H-1B, you can apply for another job in the United States, and you will be permitted to work for that new employer while their H-1B application for you is pending.

---

[11] United States Citizenship and Immigration Services, formerly the Department of Immigration and Naturalization Services (INS). Due to various reshuffles after the creation of the Department of Homeland Security, you will find this department referred to almost interchangeably as USCIS, BCIS (Bureau of Citizenship and Immigration Services), or INS.

As you can see, the process is structured in such a way as to avoid depressing local wages, and to give employment to U.S. residents preferentially before authorizing an imported worker to fill the position. A few laws have also been tweaked—see particularly my last point—to avoid some of the stress this process causes on the H-1B holder. Being an H-1B, especially if your employer isn't doing very well financially, is very frightening.

That's how it all works in theory; there are various abuses that are reported to occur. The most commonly reported abuse is that shell companies will bring H-1B talent in at a low rate (based on prevailing wages wherever the shell company is based) and hire them out to companies located elsewhere at a massive hourly rate. Sometimes the H-1B is not even paid the amount that was specified in the LCA. The H-1B is paid a pittance; the shell company pockets the difference. Purportedly, these companies also treat their employees virtually like slaves; they're threatened with withdrawal of sponsorship (which means an end to their eligibility for U.S. employment) if they complain.

At the end of the day, I can judge the availability of embedded jobs in America largely by the recruitment activity I see, and that activity is formidable and growing. Ignore the doom and gloom; outsourcing and worker visas are not going to preclude you from finding work in the United States. I would, however, advise you to keep ascending the management track if this option is offered to you; outsourcing *does* occur, and it's usually happening to "individual contributors." On the other hand, the need for competent project managers is growing enormously. Being able to assume either role makes it easier for you to avoid the need to look for a job with a new company if your engineering role is outsourced.

7.4 Procedures and You: Keeping Your Head Above Water

At least in America, few (if any) large corporations are the monolithic entities they pretend to be. Although they may be unified under a single brand name, with a single corporate philosophy on paper, in practice, most large corporations are like Chinese dragons at the lunar new year. In other words, they're a bunch of potentially self-sufficient individual units draped in something opaque to make them look like a single organism. These units don't necessarily share a common philosophy, and they usually have a widely disparate array of internal procedures, some of which may be fundamentally incompatible with the "big picture" methods and goals of the parent corporation.

This situation is pretty much unavoidable; it's extremely expensive (in terms of cash and customer satisfaction, among other things) just to dump an entirely new set of procedures and goals on every employee of a newly acquired subsidiary and force an overnight switchover. In fact, the cost is approximately the same as it would be to fire and rehire everyone from the subsidiary into the parent company.

In a large-corporation job, you're inevitably going to be exposed to many more procedural requirements than you would be in a small company. Speaking from the worm's-eye view of a frontline engineer, we can break these procedures down into the following broad categories.

- *Type 1*: Tasks which are clearly necessary or useful in order to perform engineering functions effectively; for example, a standardized process for handling electronic CAD symbols, procedures for issuing engineering change notifications to existing products, and so on.

- *Type 2*: Tasks which are not, strictly speaking, engineering necessities, but are practically important in some way and hence can't be ignored. ISO procedures, government required test data, etc. fall into this category. Almost all of these tasks are directly a result of customer requirements, and hence have an unambiguous bottom-line effect.

- *Type 3*: Tasks which are designed to generate metrics or other outputs that don't directly impact engineering functions in any way. Sometimes, these metrics exist only so that engineering functions can be measured using the same units as, say, marketing. This is a bit like trying to measure the acidity of orange juice in inches.

(There's also a fourth category of tasks—things that are really somebody else's job but somehow seem to wind up becoming your responsibility. These are obviously the most irritating of all.)

Tasks of Types 1 and 2 are simply things you're going to have to live with; they're part of the job you're taking and although you might tactfully suggest improvements where appropriate,[12] at the end of the day you're just going to have to knuckle under and do the legwork to satisfy the requirements of these procedures.

The third category is the interesting one. Generally, most of your "wasted" time is going to be spent on these tasks. Engineers almost always have a strong antipathy toward these sorts of jobs, and with excellent reason—they don't generate "useful" output and the only reason you're required to do them is because they provide a means of measuring your compliance with company policies. As a side note, many waste tasks are generated by the need ISO9000 compliance. It isn't actually difficult to be ISO900x compliant—all you really need to do is document a process, and prove that you follow it.[13] It's nominally a compliant system if you say "Our process for forecasting sales is to throw chicken bones on a calculator keypad," as long as you can prove that you actually did it for projects in progress.

The problem with ISO9000 is that somebody in management will want to see some piece of paperwork generated for a particular step in the development cycle. In order to ensure that this piece of paperwork is generated, they'll add it to the list of ISO9000 deliverables. Much like a body of laws, the list of ISO9000 deliverables rarely shrinks—so the burden of keeping the paperwork up to date expands relentlessly.

So, how do these "waste" tasks insinuate themselves into the engineering process? From personal observation, a large proportion of these tasks derive from an inappropriate application of management tools. For example, there are some very reliable, well-established techniques for analyzing the outputs of an ongoing process and using this information to adjust input parameters or process steps in order to improve output quality. These tools break down for many engineering projects, because, at least in theory, engineering's output is a design that is bounded entirely by the product specification. In other words, marketing says "Design us a widget that generates ten megathingies of output and lasts five years," and once engineering has achieved that, their task is nominally over. There is no

---

[12] Needless to say, your first day on the job is probably not a good time for such suggestions.

[13] The purpose of this certification is to establish that you have a documented way of doing business. It really has nothing to do with quality per se.

ongoing process that can be tuned to improve that specific result (although the product can, of course, be improved, that would be a separate project. Obviously there are also manufacturing improvements that could be made, but again these aren't in the scope of the original engineering task).

Before I get thousands of complaint letters, please note an important semantic point: Process tools can, with perfect validity, be used to analyze and improve the process of developing products (including the engineering phase). They just don't work very well for a large percentage of individual engineering projects, particularly new product development. Anything that has a defined endpoint doesn't lend itself to effective analysis with tools oriented toward ongoing processes. Imagine the MBAs at Ford looking at a defective car coming off the production line and trying to develop a procedure that would magically go back in time and retroactively fix that particular car, and you've got an idea of what I'm talking about.

Another large chunk of Type 3 tasks originate from company-global technologies that aren't appropriate for all types of engineering. The way this usually comes about is that one slice of the business introduces a new technology and shows a big productivity increase. Managers of the other business units obviously want similar increases, so they import the technology into totally unrelated fields. This isn't always a resounding success; Procrustean[14] techniques are usually necessary and the result is often appropriately painful.

As a prime example of this, compare the engineering flow of designing a car door handle versus designing a radio receiver. The door handle can be modeled with near-perfect accuracy in software. Knowing what kind of resin will be used to cast the part, the mechanical engineer can generate extremely reliable estimates of weight, strength and other important parameters. They can also put the virtual part into a functional 3D software model of the car and test it for fit and function without even spending a dime on stereolithographic prototypes or other tangibles. On the other hand, once the engineer declares "This part is

---

[14] Procrustes was a legendary inkeeper, of sorts. He would hospitably invite people to stay the night at his house, but if they were the wrong length for the bed he would either stretch them on the rack or cut off their feet so they would fit exactly. The adjective Procrustean is sometimes used in computer science when talking about string-handling functions. Procrustes was given a taste of his own bed-making technique by Theseus, a hero whose achievements many engineers aspire to emulate.

ɔod," that instantly triggers an expense of several tens of thousands of dollars to make an injection mold. Engineering departments designed around these sorts of tasks—easy to simulate, expensive to prototype—tend to have complex, multilevel approval processes with a distinctly one-way flow. Each step will generate required paperwork—often, detailed simulation results that verify specific properties of the part being designed. The assumption, based on the high cost of going to the prototype stage, is that engineering time inspecting, reinspecting and approving drawings and simulations is always less costly than wasted tooling. In these sorts of systems, it is usually a difficult special-case event to demote a design from "finished" to "design work in progress" if rework is required due to problems discovered with the prototype. It's generally also impossible to bypass irrelevant process steps if you do somehow manage to demote a design.

Building a radio receiver is a very different sort of problem. Although there is software commercially available that can model the various effects at work on the circuit board, it's fantastically difficult to include all the required variables and as a result, it's a time-consuming and often inaccurate job to generate simulations. Trying to simulate a product of this kind to the level of accuracy necessary to make irrevocable design decisions is amazingly difficult. As a result, the typical process for developing a product like this is as follows:

1. The circuit design engineers develop a schematic. Where possible, this will often consist of pre-verified design elements borrowed from other products. Individual elements may be simulated at a circuit level.

2. The schematic is reviewed and approved by the entire engineering team. Typically the only required approval will be the project leader, since the tooling for this project is not a major capital investment.

3. The schematic passes to the PCB layout team. Accompanying the schematic will be a list of verbal rules specific to this particular product; e.g., "Keep U5 and Q1 as far away as possible" or "Ensure signals D0-D7 are routed far away from U12."

4. The PCB layout team does a first cut at the layout, and passes a (paper) copy back to the engineering team.

5. The engineering team will inspect the critical areas of the layout. If sub-optimal copper structures are noticed, these will be bounced back to the PCB layout team and the process will iterate again. Usually one or two passes, at most, will resolve all the obvious issues.

6. The PCB engineering department will then perform a manufacturability review on the design, to ensure that the component layout is compatible with the assembly process being used. This step rarely requires further input from the circuit design engineers.

7. Prototype PCBs are ordered. The cost of a small validation prototype run is typically well under $1,000 in cash terms, but may be somewhat higher than this depending on special circumstances such as the urgency of the order, the PCB vendor's panelization rules and the requirements of the equipment used to place and solder components onto the board.

8. The RF engineer performs laboratory and field tests on the prototypes to determine receiver performance. This step almost always involves tweaking some component values.

9. If performance is adequate with the first-cut PCB layout, the process is complete. Otherwise, the circuit design team will sit down with the layout and brainstorm it again. The entire process iterates back to the third step, as shown above.

Is this technique efficient? An approximate ballpark figure for engineering time in the United States is $100/hour.[15] This number is obviously very rough, but it's useful for simple analyses like this. If we estimate that our PCB prototype run will cost $1,000 (equivalent to 1.25 person-days of engineering time), and the associated layout tweaking and performance testing will take three person-days, we have a total cost of 4.25 person-days, or a cash equivalent of $3,400 per design spin. Compare this to the fact that you could easily spend a week

---

[15] This is a crude budgetary number, reflecting the miscellaneous costs of having a person on payroll; health insurance, payroll taxes, electricity, IT support, toilet paper, coffee and numerous other small factors are built into this figure. The actual salary of the engineer will sometimes be about 50% of this number, often less.

or two weeks building a really accurate software simulation of the design (even assuming you can gather all the parameters), and $3,400 plus some risk looks like rather good value.

Back to the question at hand, which is to survive and prosper in your new job. For the moment, I'll assume that your primary goal is to be a good employee and march up the normal raise-and-promotion ladder in your chosen company. Clearly, you want to do as much useful work as possible. Since Type 3 tasks steal time that you could otherwise be using to perform that useful work, you obviously want to minimize the number of Type 3 tasks that you carry out. Here's how to do that and still look like a friend of the corporation. Considering only Type 3 tasks, first carry out these steps:

1. Establish on what you are going to be assessed.

2. Establish by whom you are going to be assessed.

3. Determine how much personal buy-in your assessor has for each procedure listed in step 1.

4. Find out if any of your required outputs are already being generated by other people.

Step 1 is fairly obvious. If nobody is ever going to look at your output for a given task, there is no good reason to do that task. Quite possibly the process evangelists at your company will faint if they hear that you're skipping a task with some religious significance to the corporate mantra, but this really isn't your problem. The key here—and I'm not being flippant when I say this—is to minimize your contact with those evangelists, and hence minimize the amount of time they spend scrutinizing how you dot your *i*'s and cross your *t*'s.

Step 2 is also obvious. Your assessor is most likely your immediate supervisor. In cases where this isn't true, you should be able to identify the relevant person quite readily.

Step 3 is a tricky one, and can only be carried out successfully by maintaining a good two-way dialog with your assessor. The reason you need to know how much buy-in your assessor has is because you need to know how much enthusiasm to show when you're carrying out these waste tasks. If your supervisor is a staunch supporter of the time-wasting process you're trying to avoid, you need to be diligent. However, if they don't have a strong personal interest in this piece of paperwork (or whatever it is), then you can safely ascertain the bare minimum requirements and do just that, nothing more.

Possibly the worst case you have to deal with is where the official corporate manual tells you to use procedure ABC, but your assessor actively discourages that procedure in favor of some other technique XYZ. This XYZ is typically a "legacy" procedure left over from your business unit's preacquisition life as an independent corporation. If your supervisor absolutely refuses to let you do things the "corporate way," you're in a sticky situation. Although you're potentially covered by the argument that you were ordered to do things in a certain way, this argument has failed spectacularly for soldiers many times in the past and could easily fail for you too. I would suggest you obey your supervisor, but email them and ask for clarification on how the way you are doing things matches with published company policy. The email is an audit trail that can be used to cover your rear if problems arise later.

Step 4 has great potential to save you time, especially for tasks that consist primarily of establishing an audit trail; ISO documents in particular. You'll probably find that most of the requirements are already being generated in the form of meeting minutes, signed authorizations to purchase tooling, and so forth. Make sure you learn where these "free" outputs are generated, and you won't need to run around chasing signatures and suchlike.

In summary: Although what I'm about to say is rank heresy, it's a view commonly held by engineers, and one to which I personally subscribe. ***Procedures, no matter how well-designed, cannot create quality; they can, at best, only provide a standardized way of measuring it.*** You cannot avoid learning and using your company's procedures, but you can try to minimize the negative work these procedures generate for you.

7.5   Managing Relationships with Marketing

The lines of communication among and division of responsibilities between departments of a large company often follow very similar patterns, no matter what the company might be.

> ### Teleology (n.)
>
> 1. The study of design or purpose in natural phenomena.
>
> 2. The use of ultimate purpose or design as a means of explaining phenomena.
>
> 3. Belief in or the perception of purposeful development toward an end, as in nature or history.
>
> —*The American Heritage Dictionary of the English Language*, Fourth Edition

Some might argue teleologically; the reason corporate behaviors evolve this way is because they are destined to evolve this way.

In almost any organization—even very small companies consisting of only two or three persons—product features, release schedules and so on are driven, or perhaps herded, by marketing. This is a normal and, believe it or not, a desirable state of affairs. Marketing people—if they're doing their job properly—have their finger on the pulse of what the customers want, when they want it, and how much they can be inveigled into paying for it. Like it or not, in a for-profit company, these factors are the primary constraints within which you, the engineer, need to shape your design. Even in a pure research environment—and there are precious few of those jobs—"marketing" pressure comes from the people who direct funding toward specific research goals.

In a hypothetical perfect company, marketing should hand down a wishlist of features. Engineering then responds, after due analysis, with reasonably accurate estimates of the development cost, development time, and per-unit costs for the various features under discussion. Marketing will forecast the likely sales for the product and present all this information to upper management, who will then make the requisite profit-and-loss decision as to whether this particular widget is profitable enough to produce. In some cases they might also assess if it's worth producing, even if not profitable, in order to achieve some other business goal.

Sony, for instance, makes a bunch of showcase gadgets that aren't necessarily intended to be profitable, simply because its corporate strategy includes a desire to be perceived as leading its competitors in innovative designs.

It's also relatively common for products to be introduced as loss leaders for either other, more profitable products, or for IP licensing deals. As an illustration of this, Sony's proprietary Memory Stick storage card format was promoted very heavily on its initial introduction; Sony was virtually paying other companies to develop products that would use the Memory Stick. Now, the shoe's on the other foot—it's relatively expensive to become part of the Memory Stick club, and presumably Sony is now starting to see actual revenue from the development effort.

Real life, of course, is considerably woollier than my highly idealized previous description. For instance, it's an unfortunate but utterly inescapable fact that many if not all engineering projects have a certain seasoning of theoretical research about them. Keep in mind that I don't use that phrase "theoretical research" in quite the way a scientist would; I use it to mean the process whereby you take a list of proposed features—usually quite vague—and decide what sort of hardware will be required to implement them. This process is a delicate balance, and is much more of an art form than a science. It seeks a solution that is bounded by the following limits:

– *How much time do we have to spend on early analysis?* The range of parts available to us is approximately infinite.[16] The more time we spend scouring the earth for new microcontrollers, ASICs and ASSPs, the more likely we are to find a "better" solution than the current best proposal.

– *How much money can we spend acquiring test equipment?* If we can actually build a quick and dirty prototype, or at least port some of our critical firmware onto a reference board for the chip under consideration, we can gauge rapidly whether the part is adequate or not. This kind of spending is scientific research funding, though—not actual product engineering.

---

[16] By this I mean that the only condition effectively limiting the resources we can spend searching is that point at which it is no longer worth continuing the search. In other words, we can find a "good" solution and continue searching for a "better" solution, but we can never find the "best" solution. It usually isn't even possible to define what the "best" solution might be, although most attempts at such a definition normally begin by assuming a hypothetical custom device that contains exactly the features we need and no more.

Upper management generally doesn't like these sorts of expenditures. There is a critical danger here, too, that the quick and dirty technology demonstration prototype may seem like a tempting foundation for the real product.

— *What is the cost of overestimating a particular part's capabilities?* In other words, what will be the consequences if we look at all the datasheets, run some calculations, and decide that part XYZ will be suitable and has an optimal price point, but it later transpires that it's just a shade under-powered for the task at hand? A large company will be able to swallow six months of wasted development time and, say, ten thousand dollars worth of platform-specific development tools. (The individual engineer who chose to go down that path might not survive, of course.)

— *What are the anticipated product volumes and margins?* This information will probably be included in the product request from marketing (though it is quite normal for it to be wildly inaccurate, particularly for a project that has a whiff of anything new and untried in it—marketing simply doesn't have the information to predict what the bill of materials cost will be). The reason these data figure into your solution set is because if you're only going to sell five hundred units at a profit of, say, ten dollars each, it's not worth spending very much time optimizing those designs.

Unfortunately, most engineers don't have a good enough intuitive grasp on the issues to estimate all this stuff accurately up front. There are many unknowns, and any off-the-cuff answer is dangerous. Worse, many engineers have a terrible propensity to make promises that can't be fulfilled.

The technique which I have found to work best for me is to limit my contact, and in fact my whole team's contact, with marketing. In particular, it is vitally important for engineers to avoid discussing speculative or forward-looking projects with marketing personnel in an unstructured way. If you have team members who can't resist talking about the latest new cool thing they're working on, these people need to be silenced or sequestered at all costs. This remark is even applicable for features you believe to be minor changes to existing projects.

In short, when dealing with marketing, I advise you to live your life by Bo Diddley's maxim: "Don't let your mouth write no check that your tail can't cash."

7.6 Task Breakdown: A Typical Week

Similarly to Section 6.6, this section breaks down a representative week in the life of an engineer working in a large company. The goal in this section is to illustrate approximately what proportion of your time you'll be spending in different tasks. I encourage you to take a look back at Section 6.6 and compare the two schedules; they illustrate quite accurately how the task mix differs between small and large company engineering positions.

- Monday
  - 2 hours – Weekly team planning meeting.
  - 1 hour – Maintaining schedules and ISO9000 paperwork for current projects.
  - 1 hour – Researching second sources for parts, validating specifications of cheaper substitute parts submitted by component engineering, and so forth.
  - 1 hour – Reviewing schematic from PCB engineering group.
  - 2 hours – Developing code and/or circuit design.
  - 1 hour – Miscellaneous (conversations with vendors, answering co-workers' questions, dealing with technical queries from other departments, and so on).
- Tuesday
  - 1 hour – Researching second sources for parts, validating specifications of cheaper substitute parts submitted by component engineering, and so forth.
  - 6 hours – Combined circuit and software development.
  - 1 hour – Miscellaneous.
- Wednesday
  - 2 hours – Weekly group/department project planning meeting.
  - 1 hour – Meeting with vendor.

- 2 hours – Auditing circuit design against regulatory standards document.

- 2 hours – Combined circuit and software development.

- 1 hour – Miscellaneous.

- Thursday

  - 1 hour – Researching second sources for parts, validating specifications of cheaper substitute parts submitted by component engineering, and so forth.

  - 1 hour – Maintaining schedules and ISO9000 documentation for current projects.

  - 2 hours – Reviewing PCB layout from PCB engineering group.

  - 2 hours – Strategy/planning meeting with marketing.

  - 1 hour – Combined circuit and software development.

  - 1 hour – Miscellaneous.

- Friday

  - 0.5 hours – Preparing timesheet for week just completed.

  - 1 hour – Dealing with manufacturing issues.

  - 2 hours – Design review for project (either software or hardware; you wouldn't normally hold both at the same time, as different teams need to be present).

  - 2 hours – Working with quality assurance on projects currently being tested.

  - 1.5 hours – Combined circuit and software development.

  - 1 hour – Miscellaneous.

8

Conclusion

Go Forth and Conquer

Congratulations—you made it to the end of the book! Hopefully, by this time you have at least the initial answers to some of the questions most frequently asked by people seeking a toehold in the embedded world. To recap where you should be after reading through this book in its entirety:

- You've had a peek at the normal educational requirements to become an embedded engineer.

- I've described how to handle nontraditional paths into the field, including information on how to survive if you don't (yet) have a degree.

- I've highlighted some of the educational paths you definitely *don't* want to take.

- Chapters 3 and 4 painted a broad-stroke picture of several popular low-end and high-end embedded architectures, and gave you some selection criteria to help you decide which languages and architectures to use when you're starting the learning process. I've tried to describe where you might want to start if you're currently an electronics guru without much programming experience, and where you might want to direct your attention if you're a computer science whiz who doesn't have much experience designing hardware and real-time software. I also illustrated a typical sort of 8-bit project and provided links to the sourcecode if you're interested in that specific circuit.

- You've had a brief introduction to the challenges, pleasures, sillinesses, and rewards of life as a freelance consultant, as an engineer (perhaps the sole engineer) in a small company, and as one of a large stable of engineers in a big company.

I'm always interested in hearing from my readers. If you have comments or questions about the material in this book—particularly if you feel I didn't explain some issue that's vital to your own situation, please feel free to email me at *sysadm@zws.com*. Questions and answers that come up frequently, or that seem to be of interest to a variety of people, will be posted in my publications support area at *<http://www.zws.com/publications/>*.

Index